# Information Society and Information Economy

# Information Society and Information Economy

Antra Arora
&
Dr. Neha Arora

RANDOM PUBLICATIONS
NEW DELHI (INDIA)

**Information Society and Information Economy**

---

ISBN 978-93-5111-507-6


Published in 2015 in India by

**RANDOM PUBLICATIONS**

4376-A/4B, Gali Murari Lal, Ansari Road
New Delhi-110 002
Phone : +9111-43580356, 011-23289044, 011-43142548
e-mail: sales@randompublications.com,
info@randompublications.com, randomexports@gmail.com

*Type Setting by* : Friends Media, Delhi-110089
*Digitally Printed at:* Replika Press Pvt. Ltd.

# Preface

Over recent years, we are become increasingly aware of the rapid advances in information and telecommunications technology and of the ways in which this has impacted on our society and economy. Digital technology has created inextricable links between telecommunications and computing technologies, transforming the ways in which information is exchanged and accessed. It has influenced the ways in which business is conducted, and how governments and their instrumentalities interact with the business community and society generally. Personal computers and the Internet are increasingly contributing to social and economic change.

The terms 'information society' and 'information economy' have been in common use for some time and are generally taken to relate to these phenomena. The term 'information society' is mainly used to refer to the diffusion of these technologies throughout the community (business, government and households) , and the term 'information economy' relates to the flow of information between economic units, the transactions that take place, and the benefits resulting from these transactions and information flows.

This eye-opening, fact-filled book profiles the rise of the net generation, which is using digital technology to change the way individuals and society interact. It presents key information relating to the information society and the information economy. It describes the information and communications technologies sector, which is the part of the economy that produces information and telecommuniucation goods and services, then it examines their use by society, businesses, farms and individuals. The book will be an essential reading for parents, teachers, policy makers, marketers, business leaders, social activists and others.

**Author**

# Contents

# 1

# The Growth of Information in Society

An information society is a society where the creation, distribution, use, integration and manipulation of information is a significant economic, political, and cultural activity. The aim of the information society is to gain competitive advantage internationally, through using information technology (IT) in a creative and productive way. The knowledge economy is its economic counterpart, whereby wealth is created through the economic exploitation of understanding. People who have the means to partake in this form of society are sometimes called digital citizens. This is one of many dozen labels that have been identified to suggest that humans are entering a new phase of society.

The markers of this rapid change may be technological, economic, occupational, spatial, cultural, or some combination of all of these. Information society is seen as the successor to industrial society. Closely related concepts are the post-industrial society), post-fordism, post-modern society, knowledge society, Telematic Society, Information Revolution, Liquid modernity, and network society.

There is currently no universally accepted concept of what exactly can be termed information society and what shall rather not so be termed. Most theoreticians agree that a transformation can be seen that started somewhere between the 1970s and today and is changing the way societies work fundamentally. Information technology is not only internet, and there are discussions about how big the influence of specific media or specific modes of production really is.

In 2005, governments reaffirmed their dedication to the foundations of the Information Society in the Tunis Commitment and outlined the basis for implementation and follow-up in the Tunis Agenda for the Information Society. In particular, the Tunis Agenda addresses the issues of financing of ICTs for development and Internet governance that could not be resolved in the first phase.

Some people, such as Antonio Negri, characterize the information society as one in which people do immaterial labour. By this, they appear to refer to the production of knowledge or cultural artifacts. One problem with this model is that it ignores the material and essentially industrial basis of the society. However it does point to a problem for workers, namely how many creative people does this society need to function? For example, it may be that you only need a few star performers, rather than a plethora of non-celebrities, as the work of those performers can be easily distributed, forcing all secondary players to the bottom of the market. It is now common for publishers to promote only their best selling authors and to try to avoid the rest—even if they still sell steadily. Films are becoming more and more judged, in terms of distribution, by their first weekend's performance, in many cases cutting out opportunity for word-of-mouth development.

Considering that metaphors and technologies of information move forward in a reciprocal relationship, we can describe some societies (especially the Japanese society) as an information society because we think of it as such as letters.

## Growth of Technologically Mediated Information

The growth of technologically mediated information has been quantified in different ways, including society's technological capacity to store information, to communicate information, and to compute information. It is estimated that the world's technological capacity to store information grew from 2.6 (optimally compressed) exabytes in 1986, which is the informational equivalent to less than one 730-MB CD-ROM per person in 1986 (539 MB per person), to 295 (optimally compressed) exabytes in 2007. This is the informational equivalent of 60 CD-ROM per person in 2007 and represents a sustained annual growth rate of some 25%. The world's combined technological capacity to receive information through one-way broadcast networks was the informational equivalent of 174 newspapers per person per day in 2007. The world's combined effective capacity to exchange information through two-way telecommunication networks was 281 petabytes of (optimally compressed) information in 1986, 471 petabytes in 1993, 2.2 (optimally compressed) exabytes in 2000, and 65 (optimally compressed) exabytes in 2007, which is the informational equivalent of 6 newspapers per person per day in 2007. The world's technological capacity to compute information with humanly guided general-purpose computers grew from 3.0 × 10^8 MIPS in 1986, to 6.4 x 10^12 MIPS in 2007, experiencing the fastest growth rate of over 60% per year during the last two decades.

James Beniger describes the necessity of information in modern society in the following way: "The need for sharply increased control that resulted from the industrialization of material processes through application of inanimate sources of energy probably accounts for the rapid development of automatic feedback technology in the early industrial period (1740-1830)" "Even with enhanced feedback control, industry could not have developed without the enhanced means to process matter and energy, not only as inputs of the raw materials of production but also as outputs distributed to final consumption."

## Development of the Information Society Model

One of the first people to develop the concept of the information society was the economist Fritz Machlup. In 1933, Fritz Machlup began studying the effect of patents on research. His work culminated in the study "The production and distribution of knowledge in the United States" in 1962. This book was widely regarded and was eventually translated into Russian and Japanese. The Japanese have also studied the information society.

The issue of technologies and their role in contemporary society have been discussed in the scientific literature using a range of labels and concepts. This section introduces some of them. Ideas of a knowledge or information economy, post-industrial society, postmodern society, network society, the information revolution, informational capitalism, network capitalism, and the like, have been debated over the last several decades.

Fritz Machlup introduced the concept of the knowledge industry. He distinguished five sectors of the knowledge sector: education, research and development, mass media, information technologies, information services. Based on this categorization he calculated that in 1959 29% per cent of the GNP in the USA had been produced in knowledge industries.

Peter Drucker has argued that there is a transition from an economy based on material goods to one based on knowledge. Marc Porat distinguishes a primary (information goods and services that are directly used in the production, distribution or processing of information) and a secondary sector (information services produced for internal consumption by government and non-information firms) of the information economy. Porat uses the total value added by the primary and secondary information sector to the GNP as an indicator for the information economy. The OECD has employed Porat's definition for calculating the share of the information economy in the total economy (e.g. OECD 1981, 1986). Based on such indicators, the information society has been defined as a society where more than half of the GNP is produced and more than half of the employees are active in the information economy.

For Daniel Bell the number of employees producing services and information is an indicator for the informational character of a society. "A post-industrial society is based on services. (…) What counts is not raw muscle power, or energy, but information. (…) A post industrial society is one in which the majority of those employed are not involved in the production of tangible goods".

Alain Touraine already spoke in 1971 of the post-industrial society. "The passage to postindustrial society takes place when investment results in the production of symbolic goods that modify values, needs, representations, far more than in the production of material goods or even of 'services'. Industrial society had transformed the means of production: post-industrial society changes the ends of production, that is, culture. (…) The decisive point here is that in postindustrial society all of the economic system is the object of intervention of society upon itself. That is why we can call it the programmed society, because this phrase captures its capacity to create models of management, production, organization, distribution, and consumption, so that such a society appears, at all its functional levels, as the product of an action exercised by the society itself, and not as the outcome of natural laws or cultural specificities" (Touraine 1988: 104). In the programmed society also the area of cultural reproduction including aspects such as information, consumption, health, research, education would be industrialized. That modern society is increasing its capacity to act upon itself means for Touraine that society is reinvesting ever larger parts of production and so produces and transforms itself. This makes Touraine's concept substantially different from that of Daniel Bell who focused on the capacity to process and generate information for efficient society functioning.

Jean-François Lyotard has argued that "knowledge has become the principle [sic] force of production over the last few decades". Knowledge would be transformed into a commodity. Lyotard says that postindustrial society makes knowledge accessible to the layman because knowledge and information technologies would diffuse into society and break up Grand Narratives of centralized structures and groups. Lyotard denotes these changing circumstances as postmodern condition or postmodern society.

Similarly to Bell, Peter Otto and Philipp Sonntag say that an information society is a society where the majority of employees work in information jobs, i.e. they have to deal more with information, signals, symbols, and images than with energy and matter. Radovan Richta argues that society has been transformed into a scientific civilization based on services, education, and creative activities. This transformation would be the result of a scientific-technological transformation based on technological progress and the increasing importance of computer technology. Science and technology would become immediate forces of production.

Nico Stehr says that in the knowledge society a majority of jobs involves working with knowledge. "Contemporary society may be described as a knowledge society based on the extensive penetration of all its spheres of life and institutions by scientific and technological knowledge". For Stehr, knowledge is a capacity for social action.

Science would become an immediate productive force, knowledge would no longer be primarily embodied in machines, but already appropriated nature that represents knowledge would be rearranged according to certain designs and programs. For Stehr, the economy of a knowledge society is largely driven not by material inputs, but by symbolic or knowledge-based inputs, there would be a large number of professions that involve working with knowledge, and a declining number of jobs that demand low cognitive skills as well as in manufacturing.

Also Alvin Toffler argues that knowledge is the central resource in the economy of the information society: "In a Third Wave economy, the central resource – a single word broadly encompassing data, information, images, symbols, culture, ideology, and values – is actionable knowledge".

At the end of the twentieth century, the concept of the network society gained importance in information society theory. For Manuel Castells, network logic is besides information, pervasiveness, flexibility, and convergence a central feature of the information technology paradigm. "One of the key features of informational society is the networking logic of its basic structure, which explains the use of the concept of 'network society'". "As an historical trend, dominant functions and processes in the Information Age are increasingly organized around networks. Networks constitute the new social morphology of our societies, and the diffusion of networking logic substantially modifies the operation and outcomes in processes of production, experience, power, and culture".

For Castells the network society is the result of informationalism, a new technological paradigm. Jan Van Dijk defines the network society as a "social formation with an infrastructure of social and media networks enabling its prime mode of organization at all levels (individual, group/ organizational and societal). Increasingly, these networks link all units or parts of this formation (individuals, groups and organizations)".

For Van Dijk networks have become the nervous system of society, whereas Castells links the concept of the network society to capitalist transformation, Van Dijk sees it as the logical result of the increasing widening and thickening of networks in nature and society. Darin Barney uses the term for characterizing societies that exhibit two fundamental characteristics: "The first is the presence in those societies of sophisticated – almost exclusively digital – technologies of networked communication and information management/distribution, technologies which form the basic infrastructure mediating an increasing array of social, political and economic practices. (…) The second, arguably more intriguing, characteristic of network societies is the reproduction and institutionalization throughout (and between) those societies of networks as the basic form of human organization and relationship across a wide range of social, political and economic configurations and associations".

The major critique of concepts such as information society, knowledge society, network society, postmodern society, postindustrial society, etc. that has mainly been voiced by critical scholars is that they create the impression that we have entered a completely new type of society. "If there is just more information then it is hard to understand why anyone should suggest that we have before us something radically new". Critics such as Frank Webster argue that these approaches stress discontinuity, as if contemporary society had nothing in common with society as it was 100 or 150 years ago. Such assumptions would have ideological character because they would fit with the view that we can do nothing about change and have to adopt to existing political realities. These critics argue that contemporary society first of all is still a capitalist society oriented towards accumulating economic, political, and cultural capital. They acknowledge that information society theories stress some important new qualities of society (notably globalization and informatization), but charge that they fail to show that these are attributes of overall capitalist structures. Critics such as Webster insist on the continuities that characterise change. In this way Webster distinguishes between different epochs of capitalism: laissez-faire capitalism of the 19th century, corporate capitalism in the 20th century, and informational capitalism for the 21st centur.

For describing contemporary society based on a dialectic of the old and the new, continuity and discontinuity, other critical scholars have suggested several terms like:

- *transnational network capitalism, transnational informational capitalism*: "Computer networks are the technological foundation that has allowed the emergence of global network capitalism, that is, regimes of accumulation, regulation, and discipline that are helping to increasingly base the accumulation of economic, political, and cultural capital on transnational network organizations that make use of cyberspace and other new technologies for global coordination and communication. [...] The need to find new strategies for executing corporate and political domination has resulted in a restructuration of capitalism that is characterized by the emergence of transnational, networked spaces in the economic, political, and cultural system and has been mediated by cyberspace as a tool of global coordination and communication. Economic, political, and cultural space have been restructured; they have become more fluid and dynamic, have enlarged their borders to a transnational scale, and handle the inclusion and exclusion of nodes in flexible ways. These networks are complex due to the high number of nodes (individuals, enterprises, teams, political actors, etc.) that can be involved and the high speed at which a high number of resources is produced and transported within them. But global network capitalism is based on structural inequalities; it is made up of segmented spaces in which central hubs (transnational corporations, certain political actors, regions, countries, Western lifestyles, and worldviews) centralize the production, control, and flows of economic, political, and cultural capital (property, power, definition capacities). This segmentation is an expression of the overall competitive character of contemporary society."

- *digital capitalism*: "networks are directly generalizing the social and cultural range of the capitalist economy as never before"
- *virtual capitalism*: the "combination of marketing and the new information technology will enable certain firms to obtain higher profit margins and larger market shares, and will thereby promote greater concentration and centralization of capital"
- *high-tech capitalism or informatic capitalism* – to focus on the computer as a guiding technology that has transformed the productive forces of capitalism and has enabled a globalized economy.

Other scholars prefer to speak of information capitalism or informational capitalism. Manuel Castells sees informationalism as a new technological paradigm (he speaks of a mode of development) characterized by "information generation, processing, and transmission" that have become "the fundamental sources of productivity and power". The "most decisive historical factor accelerating, channelling and shaping the information technology paradigm, and inducing its associated social forms, was/is the process of capitalist restructuring undertaken since the 1980s, so that the new techno-economic system can be adequately characterized as informational capitalism". Castells has added to theories of the information society the idea that in contemporary society dominant functions and processes are increasingly organized around networks that constitute the new social morphology of society. Nicholas Garnham is critical of Castells and argues that the latter's account is technologically determinist because Castells points out that his approach is based on a dialectic of technology and society in which technology embodies society and society uses technology. But Castells also makes clear that the rise of a new "mode of development" is shaped by capitalist production, i.e. by society, which implies that technology isn't the only driving force of society.

Antonio Negri and Michael Hardt argue that contemporary society is an Empire that is characterized by a singular global logic of capitalist domination that is based on immaterial labour. With the concept of immaterial labour Negri and Hardt introduce ideas of information society discourse into their Marxist account of contemporary capitalism. Immaterial labour would be labour "that creates immaterial products, such as knowledge, information, communication, a relationship, or an emotional response", or services, cultural products, knowledge. There would be two forms: intellectual labour that produces ideas, symbols, codes, texts, linguistic figures, images, etc.; and affective labour that produces and manipulates affects such as a feeling of ease, well-being, satisfaction, excitement, passion, joy, sadness, etc.

Overall, neo-Marxist accounts of the information society have in common that they stress that knowledge, information technologies, and computer networks have played a role in the restructuration and globalization of capitalism and the emergence of a flexible regime of accumulation. They warn that new technologies are embedded into societal antagonisms that cause structural unemployment, rising poverty, social exclusion, the deregulation of the welfare state and of labour rights, the lowering of wages, welfare, etc.

Concepts such as knowledge society, information society, network society, informational capitalism, postindustrial society, transnational network capitalism, postmodern society, etc. show that there is a vivid discussion in contemporary sociology on the character of contemporary society and the role that technologies, information, communication, and co-operation play in it.Information society theory discusses the role of information and information technology in society, the question which key concepts shall be used for characterizing contemporary society, and how to define such concepts. It has become a specific branch of contemporary sociology.

## Second and Third Nature

An information society is the means of getting information from one place to another. As technology has become more advanced over time so too has the way we have adapted in sharing this information with each other.

"Second nature" refers a group of experiences that get made over by culture. They then get remade into something else that can then take on a new meaning. As a society we transform this process so it becomes something natural to us, i.e. second nature. So, by following a particular pattern created by culture we are able to recognise how we use and move information in different ways. From sharing information via different time zones (such as talking online) to information ending up in a different location (sending a letter overseas) this has all become a habitual process that we as a society take for granted.

However, through the process of sharing information vectors have enabled us to spread information even further. Through the use of these vectors information is able to move and then separate from the initial things that enabled them to move. From here, something called "third nature" has developed. An extension of second nature, third nature is in control of second nature. It expands on what second nature is limited by. It has the ability to mould information in new and different ways. So, third nature is able to 'speed up, proliferate, divide, mutate, and beam in on us from else where. It aims to create a balance between the boundaries of space and time. This can be seen through the telegraph, it was the first successful technology that could send and receive information faster than a human being could move an object. As a result different vectors of people have the ability to not only shape culture but create new possibilities that will ultimately shape society.

Therefore, through the use of second nature and third nature society is able to use and explore new vectors of possibility where information can be moulded to create new forms of interaction.

## Sociological Uses

In sociology, informational society refers to a post-modern type of society. Theoreticians like Ulrich Beck, Anthony Giddens and Manuel Castells argue that since the 1970s a transformation from industrial society to informational society has happened on a global scale. As steam power was the technology standing behind industrial society, so information technology is seen as the catalyst for the changes in work organisation, societal structure and politics occurring in the late 20th century. In the book Future Shock, Alvin Toffler used the phrase super-industrial society to describe this type of society. Other writers and thinkers have used terms like "post-industrial society" and "post-modern industrial society" with a similar meaning.

A number of terms in current use emphasize related but different aspects of the emerging global economic order. The Information Society intends to be the most encompassing in that an economy is a subset of a society. The Information Age is somewhat limiting, in that it refers to a 30-

year period between the widespread use of computers and the knowledge economy, rather than an emerging economic order. The knowledge era is about the nature of the content, not the socioeconomic processes by which it will be traded. The computer revolution, and knowledge revolution refer to specific revolutionary transitions, rather than the end state towards which we are evolving. The Information Revolution relates with the well known terms agricultural revolution and industrial revolution.

The information economy and the knowledge economy emphasize the content or intellectual property that is being traded through an information market or knowledge market, respectively. Electronic commerce and electronic business emphasize the nature of transactions and running a business, respectively, using the Internet and World-Wide Web. The digital economy focuses on trading bits in cyberspace rather than atoms in physical space. The network economy stresses that businesses will work collectively in webs or as part of business ecosystems rather than as stand-alone units. Social networking refers to the process of collaboration on massive, global scales. The internet economy focuses on the nature of markets that are enabled by the Internet. Knowledge services and knowledge value put content into an economic context. Knowledge services integrates Knowledge management, within a Knowledge organization, that trades in a Knowledge market. In order for individuals to receive more knowledge, surveillance is used. This relates to the use of Drones as a tool in order to gather knowledge on other individuals.

Although seemingly synonymous, each term conveys more than nuances or slightly different views of the same thing. Each term represents one attribute of the likely nature of economic activity in the emerging post-industrial society. Alternatively, the new economic order will incorporate all of the above plus other attributes that have not yet fully emerged.

## Intellectual Property Considerations

One of the central paradoxes of the information society is that it makes information easily reproducible, leading to a variety of freedom/control problems relating to intellectual property. Essentially, business and capital, whose place becomes that of producing and selling information and knowledge, seems to require control over this new resource so that it can effectively be managed and sold as the basis of the information economy. However, such control can prove to be both technically and socially problematic. Technically because copy protection is often easily circumvented and socially rejected because the users and citizens of the information society can prove to be unwilling to accept such absolute commodification of the facts and information that compose their environment.

Responses to this concern range from the Digital Millennium Copyright Act in the United States (and similar legislation elsewhere) which make copy protection circumvention illegal, to the free software, open source and copyleft movements, which seek to encourage and disseminate the "freedom" of various information products (traditionally both as in "gratis" or free of cost, and liberty, as in freedom to use, explore and share).

## References

Alistair Duff (2000) *Information Society Studies.* London: Routledge.

Christian Fuchs (2008) *Internet and Society: Social Theory in the Information Age.* New York: Routledge.

Frank Webster (2002) The Information Society Revisited. In: Lievrouw, Leah A./ Livingstone, Sonia (Eds.). *Handbook of New Media.* London: Sage. pp. 255–266.

Frank Webster (2002) *Theories of the Information Society*. London: Routledge.

# 2

# Potential Uses of ICTs

Information and communications technology or information and communication technology (ICT), is often used as an extended synonym for information technology (IT), but is a more specific term that stresses the role of unified communications and the integration of telecommunications (telephone lines and wireless signals), computers as well as necessary enterprise software, middleware, storage, and audio-visual systems, which enable users to access, store, transmit, and manipulate information.

The phrase ICT had been used by academic researchers since the 1980s, but it became popular after it was used in a report to the UK government by Dennis Stevenson in 1997 and in the revised National Curriculum for England, Wales and Northern Ireland in 2000.

The term ICT is now also used to refer to the convergence of audio-visual and telephone networks with computer networks through a single cabling or link system. There are large economic incentives (huge cost savings due to elimination of the telephone network) to merge the audio-visual, building management and telephone network with the computer network system using a single unified system of cabling, signal distribution and management.

## Development of ICTs

In the 1980s, the expectation that innovations in telecommunications and computer technologies would improve industrial performance and increase economic productivity in the industrialized nations became firmly established among the leader of the developing countries. The common position that emerged was that ICTs would allow them to leapfrog over the industrialization of their economies into a post-industrial society. Countries began to launch policies and programmes to acquire a share in international satellite communications and transborder data flow networks. In many countries, however, concern arose that ICTs might entail serious social risks, such as a potential for cultural colonialism, the replacement of jobs by machines, and the erosion of individual privacy and national sovereignty. Towards the end of the 1980s these fears seemed to have abated. A new phase began in the 1990s, characterized by a very strong fear of being left behind and cut off from the emerging global digital highway system.

In many countries, the 'digital rush' is on to ensure connections with the electronic networks for trade, finance, transport and science. This phenomenon has been inspired by the obvious benefits that digital information and communication technologies - at least in principle - seem to offer. Educational facilities can be improved by providing distance learning and on-line library access. Electronic networking has also been used to improve the quality of health services by providing remote access to the best diagnostic and healing practices and cutting costs in the process. Digital technologies for remote resource sensing can provide early warning to areas vulnerable to seismic disturbances or identify land suitable for crop cultivation.

Computer technology can contribute to the development of flexible, decentralized and small-scale industrial production. Thus the competitive position of local manufacturing and service industries can be improved. In Singapore, Brazil and Hong Kong the introduction of computer-aided manufacturing (CAM) technologies has been very successful in small-scale industries. The World Commission on Environment and Development, in its report Our Common Future suggests that 'new technologies in communication, information, and process control allow the establishment of small-scale, decentralized, widely dispersed industries, thus reducing the levels of pollution and other impacts on the local environment'.

The currently available computer-communication technologies make it fairly easy for PC users around the world to create a public sphere in 'cyberspace'. Personal computers, modems and telephone lines are being used to establish new global communities. Organizations in developing countries find it increasingly possible to join these forms of horizontal, non-hierarchical exchange that have already demonstrated their ability to counter censorship and misinformation. Groups of all types, from ecological movements, human rights activists, farmers, senior citizens, to the Zapatistas and the groups attending the Women's summit in Beijing, have made impressive use of the new, fast, reliable and effective networks of communication. The combination of telecommunication technologies with desktop publishing software has created new opportunities for even the smallest action group to disseminate its messages across the globe with relative ease and at minimal expense.

In the late 1980s and early 1990s, various aid projects introduced electronic information systems into rural health services (India), agricultural extension projects (Peru), infrastructural development (Pakistan), or energy management (Malaysia). In some developing countries, special computer training courses have been developed in schools and colleges (Sri Lanka). The growing ICT-demand in developing countries can be seen in the long waiting lists for telephone connections, the increase in cellular systems and the rapidly expanding numbers of Internet users. To meet this demand, more and more developing countries have placed the ICTs at the centre of national agendas for social and economic development. There are plans in many developing nations to enhance telecommunications infrastructures in order to facilitate participation in world markets.

The tendency to use new technologies to meet commercial objectives is problematic for people and regions without the necessary financial resources to attract companies to fulfil their needs. The challenge is made harder by the need for stable telecommunication links and electricity to power nearly all the new technologies. Both are in short supply in the poorest regions. The case of developing countries illustrates the way in which technology evolves and is diffused at different speeds across the world. Whereas new technologies are represented by state-of-the-art digital innovations in the developed world, for parts of the developing world telephones, television or even radio may be the latest technologies. It is in this context that two recent innovations in radio broadcasting represent important advances in the medium. The clockwork (or wind-up) radio is the first.

Communication specialists in development have been aware of the problem faced by isolated communities in obtaining dry-cell batteries to power radio receivers for several decades. Crystal-sets and solar-powered receivers have not proved to be entirely successful. The clockwork radio is powered by a patented power source, which with a few minutes of winding-up will keep a receiver playing for about an hour. It is an exciting invention which has been adapted to work on lap-top computers and other devices, which previously depended on batteries for electrical power.

For the present, the relatively high-costs of these radio sets prevent their extensive diffusion across the developing world. The other innovation is at the broadcaster's end. UNESCO has taken the lead in helping to develop affordable low-power radio transmitters for operating community broadcasting services. These transmitters are assembled from generic components by local technicians, who also assist with maintenance. Such services facilite the decentralization of media control to the community. They help promote participatory communication, which empowers communities, and support the planning and implementation of sustainable development programmes. The main hindrance to the spread of this technology is legislative.

Most developing countries discourage community broadcasting by limiting access to radio frequencies and to a broadcasting licence. A classic example of a new technology converging to support an old predecessor is the technology to stream sound over the Internet. This is the developed world's version of the developing world's evolution towards community broadcasting. The technology is very affordable at both ends. People can receive such Internet radio broadcasts with a relatively low investment while broadcasters can be 'on the air' broadcasting literally to the whole world at very minimal cost. These technologies represent an extremely small fraction of the total number of innovations introduced into the media industry.

The developing world with its highly limited purchasing power has sustained minimal research and development efforts to increase their access to relevant content and channels of mass communication. A study in the Philippines found that where privatization of media has been the norm, programming is not meeting the needs of local populations. This sentiment may extend quite widely across the developing world.

Transmission frequencies are a very valuable resource in some areas of the world where finding a frequency for broadcasting to minority groups is a daunting uphill battle, since regulatory authorities and established media companies have vested interests to protect. The introduction of digital television broadcasting technology with the capacity to place four channels in the same space that one analogue channel used to occupy, will soon make plenty of new broadcasting space available. A significant amount of this newly freed-up space should be reserved for minority interest broadcasting, which is now limited.

Developing countries will probably have to wait for several years before the cost of digital technologies reaches an affordable level. The early users, namely the developed countries, should therefore establish a precedent for safeguarding minority interests and those of disadvantaged groups by reserving such open digital spaces for them. Spaces must be held in trust, not only for minorities, but also for the common good in areas such as education, science and culture, which are not very profitable, but are socially vital.

## ICTs and Social Development

Across all industries, there has been a strong growth of investments in ICT applications. Spending on ICT equipment as part of overall spending on business equipment has grown dramatically in most industrial countries. In the United States, for example, such spending rose from less than 5% in 1960 to over 45% by 1996 and if the current trend continues the figure could be over 50% in 2000. According to an estimate of the United States Department of Commerce in 1998, the American ICT-industry generated $683,000 million. In some industrial countries ICT-related activities produce a growing share of the Gross Domestic Product (GDP).

Worldwide investments in telecommunication rose from $115,000 million in 1990 to $152,000 million in 1995. These investments have often been linked to the privatization of public telecommunication operators (PTOs). The privatization schemes introduced in a number of countries have raised considerable funds. The ICT industry encompasses the manufacturing of telecommu-nications equipment, computers, semiconductors and other electronic equipment, the provision of telecommunications services, computer services and software. It is the world's most important and fastest-growing industry. Of the fifty largest companies in the world, ten are ICT companies. They account for 17.5% of the total revenue of the fifty largest companies, for 23% of the total profits and for 26% of the total number of employees. In 1997, four ICT companies, General Electric, Microsoft, NTT and Intel, were among the ten largest companies in the world in terms of market value. In the same year three ICT companies, General Electric, Intel, and International Business Machines (IBM), were among the ten most profitable companies in the world. Telecommunications, semiconductors and computers are among the high growth industries in the emerging-market economies. Among the 200 top companies in these economies, twenty-two are telecommunications market operators and ten are electronics products manufacturers.

ICTs are affected by a great deal of merger activity. The largest transactions of 1998 include the SBC Communications merger with Ameritech for $272,400 million, AT&T with Telecommunications Inc. (TCI) for $48,000 million, Worldcom with MCI Comm for $37,000 million, Northern Telecom with Bay Networks for $9,000 million, and Alcatel with Digital Service Corporation (DSC) for $4,400 million. There is high degree of concentration in this industry. In the telecommunications equipment market, for example, 50% of all sales are controlled by only five companies. In the market for public telephone switching equipment, five firms (Alcatel, Siemens, Lucent, Ericsson and Nortel) control 76% of all activity.

The ICT market is growing rapidly and expanding globally. The top ten companies in the manufacturing of telecommunication equipment derive an average 61% of their revenues from sales abroad. These are Motorola (United States), Alcatel (France), Lucent (United States), Siemens (Germany), Ericsson (Sweden), NEC (Japan), Nortel (Canada), Nokia (Finland), Fujitsu (Japan), and Bosch (Germany). Most analysts expect a further growth of global sales since the domestic markets in many countries are either too small or already saturated. From various trade figures, it can be estimated that the share of the ICT industry in the world economy ranges from between 10% to 15% of total world trade. In several ICT sectors the world market shows rapid growth rates. The market for computer hardware and software, for example, grew at about 15% annually in the 1990-1995 period. In the same period, the average growth rate for total world trade was 8%. In 1997, worldwide revenues in the telecommunications industry were nearly $600,000 million. Some analysts expect this figure to grow to $1,400,000 million by the year 2000. The value of total imports in the world trade of computer equipment grew by 67% from $87,500 million in 1992 to $145,500 million in 1996. The value of total exports grew by 75% from $73,000 million in 1992 to $127,000 million in 1996 (Comtrade, United Nations Statistics Division). The market for personal computers (PCs) is still growing and may reach a sales figure of over $250,000 million in 2000.

There is also rapid growth in the world market for computer software, which in 1996 generated $109 , 3 00 million in revenues.

In the world trade of telecommunication equipment the value of total exports grew by 68% from $65,000 million in 1992 to $108,000 million in 1996. The value of total imports grew by 58% from $66,000 million in 1992 to $104,000 million in 1996. Growth rates in this sector are linked to the worldwide increase in the number of telephone lines. The growth from 519 million lines in 1990 to 693 million lines in 1995 represents an average growth rate of 6.8% per year. In 1995, 45 million new fixed telephone lines were added and 33 million people became new subscribers to mobile communications. In fact, the strongest growth rates are in the mobile communication market, particularly in the United States and Japan. In the OECD countries in 1996-67, the contribution of mobile communications to overall telecommunication services revenues amounted to some 12% and the number of subscribers doubled. In 1990, the worldwide trading of telecommunications services generated some $377,000 million. This figure increased in 1996 to over $600,000 million.

According to data from the International Telecomunications Union (ITU), the developed countries represent some 70% of overall telecommunication equipment exports, showing a $9 billion trade surplus. Particularly striking is the shift from a trade deficit for the USA in 1990 to a trade surplus of $3.2 billion in 1995. Over 100 developing countries produce no exports at all. They are obliged to import and thus show trade deficits that in 1995 exceeded $10,000 million. For most developing countries, the obvious problem is lack of capital for imports, since low overall income means that local telecommunications generate minimal revenues. The profits in national telecommunications operation are, without exception, found in international traffic, but the financial gains from overseas services do not benefit the national telecommunications carrier. For political and other reasons, many countries decide not to reinvest a sufficient share of this income in the expansion and upgrading of telecommunication facilities. Moreover, there is growing competition in the provision of international telecommunications services and strong pressures to lower rates on busy routes. Foreign investments in telecommunications have been stimulated by the privatization of former national public telecommunication operators as part of the worldwide trend toward deregulation that emerged during the 1980s. In the 1984-96 period the 44 PTOs that were privatized raised in total a sum of $159,000 million. Almost one third of the financing came from foreign investments.

## Economic Importance of ICTs

The increasing economic importance of ICTs is also linked to the growth of electronic commerce. Although for the moment electronic commerce is still a factor of limited economic significance, with revenues of approximately $26,000 million in 1998, this figure is expected to exceed the trillion dollar mark in the early 21st century. Since it is difficult to measure precisely the economic significance of electronic commerce. Several of its important features, such as easy access to data, cannot be quantified, and reliable and comprehensive statistics are not yet available; this uncertainty leads to widely varying estimates. The OECD definition of electronic commerce refers to commercial transactions that take place through open networks, like the Internet. These transactions are both business to business and business to consumers. The growing number of people around the world connected to the Internet have begun to develop a digital economy.

In cyberspace, consumers can purchase flowers, shirts, jeans, books, compact disks (CDs), tickets, hotel reservations, skin care products, pornography, kitchen equipment and household consumer goods. World Wide Web sales grew approximately from $8,000 million in 1994 to over $45,000 million in 1998. According to various analysts in Business Week, digital trade could grow from $1,200 million in 1998 to $65,000 million over the next three years. Other forecasters propose even higher figures, for instance in the travel industry, book and music sales and financial services. Electronic sales of software also hold enormous growth potential. Whatever the precise significance and validity of the forecasts, the general expectation predicts formidable growth in this sector of the global economy. Most analysts observe that this will be mainly business-to-business trading. Expectations of growth in the business-to-consumer trade are based on the assumption that on-line trading will lower production, and thus distribution, costs by up to 10% of total sales. Some 50% of this reduction of costs may be used to lower consumer prices thus stimulating further purchases by consumers.

Electronic trading via the Internet is in fact the successor to the Electronic Data Interchange (EDI) that is already used by many companies.

EDI is the exchange of electronic documents in goods trading and financial services and usually operates within closed Value Added Networks (VANs). EDI is expensive because complicated software applications have to be installed for each new trading partner and flexibility is limited because invoices and orders must be exchanged in fixed standard formats. The Internet facilitates the exchange of electronic data by accepting many different formats. However, many of the big EDI users see EDI as more reliable than current Internet traffic. In the United States at present, the volume of EDI-traffic is some fourteen times greater than that of Internet traffic. Many analysts expect that in the years ahead this ratio will change, first to a more equal distribution between EDI and Internet transactions and later to a definite advance for the Internet. This situation will be reinforced by the rate at which companies begin to share their 'intranets' with more trading partners and thus create 'extranets'.

An increasingly important part of all commercial activity on the digital market is geared toward the control over access to cyberspace. The gates that provide access to the World Wide Web - the so-called web-portals - have become crucial targets in the struggle for dominance in this market. In 1998 a veritable 'portal-fever' emerged as more and more companies sought to secure a piece of the expected profits. In June 1998 the media giant Walt Disney announced it would spend $70 million for the purchase of 43% of the stock in the search engine Infoseek. In 1998, the American National Broadcasting Company (NBC) invested $165 million for the purchase of the online service Snap! from van Cnet Inc. Time-Warner and News Corp. began to develop their own web portals.

Manufacturers of personal computers, vendors of operating systems, producers of browsers, telecommunications operators, Internet Service Providers, search engines, the big producers of information and entertainment are among the many contenders in this field. Recent acquisitions or investments among the various companies involved in cyberspace illustrate the high level of competition in this area. In 1997, Microsoft bought a 5% participation in Apple Computer for $150 million. As part of the deal, Microsoft insisted that Apple Macintosh computers be delivered with the Microsoft 'browser', Explorer. In 1998 Compaq Computer acquired Digital Equipment in a deal valued at $9,600 million. The combined revenues are estimated at around $38,000 million. WorldCom controls 60% of worldwide Internet traffic and can be seen as a major gatekeeper of access to lines and networks. In September 1997, America Online (AOL) acquired Compuserve through a complicated deal that made WorldCom the owner of Compuserve's infrastructure. WorldCom paid $1,200 million to H&R Block (Columbus, Ohio) for the purchase of Compuserve. The subscribers to Compuserve were sold to AOL. Through this deal AOL became the largest ISP on the European market with 1.5 million subscribers. On the world market, AOL has some 12 million subscribers.

The ISPs provide access to the services on the Internet and exercise considerable control over access to cyberspace. Some of them conclude agreements with the search engines. Provider MCI, for example, concluded a deal with Yahoo! in order to guide clients to the Web page of this particular search engine. The world market of ISPs is concentrated. If interest in the Web continues to grow, ISPs are likely to allow certain websites to purchase a right of 'priority'. This would mean that if too many clients browse the Web at one time, customers of the privileged sites will have priority and have to wait less time than others. There is also market concentration in this area and many linkages among the players.

There are indications of a real ICT contribution to overall economic growth, although no definite assessment is possible. In his testimony to Congress, US Federal Reserve Board Chairman Alan Greenspan pointed to increasing growth rates in productivity for the American economy and stated that: 'The dramatic improvements in computing power and communication and information technology appear to have been a major force behind this beneficial trend'. Other analysts raise serious doubts about such a causal link between ICT power and productivity. Of the world's total investment in capital goods, 50% concerns computers and peripherals. Yet expected growth in productivity has not materialized. During the period 1970-90 white collar productivity remained stagnant at around 0.9% per year and improved in the early 1990s to 1.3% - a low figure despite the announcement by Fortune magazine that 'the productivity payoff arrives'. It is not clear at present what factors cause this ICT-paradox. One possible explanation refers to the criteria that are used for the measurement of economic productivity. Another factor may also be the inadequate way in which organizations adapt to the use of ICTs. It could also be that what looks like an unexplainable delay is in fact the normal time lag needed before new technologies can lead to higher levels of productivity.

ICTs are 'synergetic' technologies and their growth therefore leads to growth in other sectors of the economy. They create an infrastructure around their products and services, similar to that of car technology earlier in the century. As with the transition from manual power to mechanization techniques and later to electromechanical innovations, today's shift towards the pervasive application of ICTs has given rise to a range of new industries such as software production, processing services, time-sharing facilities, semiconductor manufacturing, database management or electronic publishing. This potential for new economic productivity requires an educational infrastructure that teaches the knowledge and skills required for ICT-related occupations. Although there has been a considerable increase in ICT teaching in many schools and universities around the world, new multimedia tools are still widely underused or used only to supplement conventional teaching methods. There is also a lack of solid and creative training materials that cater for the specific needs of the developing countries.

The economic benefits that may accrue from the development and the deployment of ICTs are unequally distributed throughout the world. There seems to be general agreement in scientific literature and in public policy statements that the ICT gap between the developed and developing countries is widening and will be a major obstacle to the integration of all countries into the so-called Global Information Society.

The seriousness of the ICT gap is clearly demonstrated by figures on the world distribution of telephones. In 1996 there were 743.66 million main telephone lines in the world. Europe (274.23 million), the United States (170. 5 7 million) and Japan (61.53 million) represent 68% of this total, compared with 1.8% in Africa. The density of telephone lines per 100 inhabitants is also very unequally distributed. Whereas the world average is 12.88 lines per 100 inhabitants, Europe provides 34.6, the United States 63.9, Japan 48.9 and Africa 1.85. In early 1997, some 62% of the world's main telephone lines were installed in 23 affluent countries, which account for only 15% of the world's population. Over 950 million households in the world (65% of the total) were equipped with a telephone. Another indicator of this gap is the revenue from telecommunication services, which reached a world total of $620,000 million in 1996. Europe, the United States and Japan benefited from 77% of these revenues, while Africa received a mere 1.5%. Investments in the telecommunication sector show a similar distribution. In 1996 the world total was worth $166,000 million. Europe, the United States and Japan are responsible for 67% of these investments and Africa for 1.7%.

Total ICT equipment installation reveals highly uneven geographical distribution. The estimated number of PCs in the world in 1996 stood at 234,200,000. The share for Europe (72,864,000), the United States (96,600,000) and Japan (16,100,000) was 79%, while for Africa it was only 1.3%, representing 0. 6 4 computers per 100 inhabitants. The world average for that year was 4.6 computers per 10 0 inhabitants. Far above this average are the figures for Europe, 9.6, the United States, 36.2, and Japan, 12.8. Of all the 47,972,000 fax machines operating in the world in 1996, Europe was using 10,942,500, the United States 17,000,000, and Japan 14,300,000. These figures combined represent 88%, in contrast with the 0.5% in use in Africa. Of the total number of television sets in the world in 1996, Europe, the United States and Japan possessed 47% while Africa possessed only 3%.

At the present time the worldwide distribution of ICT resources is enormously unequal. In terms of availability, accessibility, and affordability of equipment and services as well as the mastery of technical and managerial skills, there are great disparities not only between affluent and developing countries, but also among different social groups within all countries. A very problematic factor is that these disparities, rather than diminishing, are growing throughout the world.

## ICTs and Employment Creation

The creation of new employment is often related to various uses of ICT, for example in new service applications in banking, shopping, education, health and business services. The development of networks combined with the lowering of communication costs has stimulated the growth of 'teleworking'. This means that people work from home and use telecommunication services. The fastest growth is generally expected in 'telecommuting' where people work alternately at home and at the office. The Internet could play a major role in creating employment for the design, maintenance and management of sites on the World Wide Web. The growth of electronic commerce initially raised concerns about implications for the labour market, but there is also the expectation that few jobs will really be lost since companies will continue to use traditional forms of commerce along with digital trading. When digital trading finally predominates, some jobs will be lost, but new ones requiring new skills will also have been created. The overall picture is characterized by contradictions. Information-related service jobs are associated with the dislocation of family and community life and threats to the health of workers, especially of women. On the other, new types of employment and modes of work organization can be highly beneficial leading to improved quality of life and greater economic resources.

There is some empirical evidence that automation eliminates jobs, but there is also evidence that automation creates new jobs. What remains uncertain is the eventual balance. The notion of a single standard impact is very inadequate and it is preferable to look at plural impacts. The skills implications of ICTs are varied and are related to differences in socio-economic, cultural and political environments. Different countries will choose different ICT-applications in order to reap different benefits from technological development. It is too early to foresee how all these factors will operate. The observation that in the affluent economies more and more employees use ICT applications in their daily work is generally uncontested. By and large, ICTs are used by some 50% of the workforce in the industrial nations.

## ICTs and Knowledge Development

One essential dimension of human development is knowledge, crucial in enabling people to broaden their choices. ICTs are the basic tools of emerging knowledge-based societies. They represent an important shift from the utilization of natural and material resources to the deployment of data and information and related analytical and processing skills. This development is, however, accompanied by national information and informatics policies through guidelines, workshops and technical assistance. Special consideration is given to 'info-ethical' questions, to achieving a balance between the common good and economic imperatives, and to facilitating the use of information and communication technologies for development purposes by the public sector and the civil society. A strong commitment by the international community is necessary to ensure that commercial interest do not dominate the development of ICTs in the world, more particularly in developing countries.

Governments, public and private institutions have to assume their responsibilities if the gap between information-rich and information-poor is not to continue to widen a strong trend toward the privatization and commercialization of knowledge sources and the concurrent enforcement of legal measures to protect private intellectual property.

The emerging global regime for Intellectual Property Rights (IPR) tends to give more emphasis to the economic aspects of IPR protection than to public interest considerations. There is a dominant economic angle in this regime which gives priority to the interests of large producers over those of small creators and consumers. The central focus is on the misappropriation of corporate property rather than on artistic and literary creativity. The move to give IPR protection to material in the public domain when it has been entered in an electronic database could render access to knowledge more costly - and thus prohibitive - for large numbers of people. In addition, the initiative to extend copyright protection to all forms of digital copying could make the Internet a pay-per-view medium creating obstacles to the access to knowledge for those who cannot pay. Developments in the international regime for intellectual property protection are an attempt to create and maintain a balance between a set of rival claims to the control over knowledge and its dissemination.

The protection of intellectual property rights must provide the incentives, rewards and recognition for individual producers of knowledge in order to stimulate progress. Benefits for the creator, and public access to artistic, literary and scientific works must both be secured. Whereas IPR protection seeks to promote the progress of science, it also restricts access to knowledge, since it defines knowledge as private property and has a tendency to facilitate monopolistic tendencies. It may well be that the current trends in the trade-oriented IPR regime lead to more restricted access rather than better public distribution. Although an absence of legal protection could have adverse effects on intellectual production, the move toward overprotection is 'not conducive to access to the networks by research workers and academics of developing countries'.

ICT developments are more technology driven than user oriented. If the ICT potential for human development is to be successfully exploited, the needs and aspirations of users must be central to the whole process of design, construction and application. The essential requirement of user orientation is hampered by the strong tendency to delegate social responsibility for the governance of the ICT-sector to a global trade regime. In fact, the politics of world communication represent a historical shift from a public service orientation to private competition with a predominantly commercial focus.

## ICTs and Cultural Development

The information and communication technologies (ICTs) are advancing with unprecedented speed, particularly in the most economically developed countries, it would be a mistake to regard them as solely responsible for the current changes. This long-heralded momentum, which seemed slow at the beginning but is now well under way, seems likely to continue at a rapid pace and, if some authors are to be believed, will usher in a new information society or information age. It is, however, one thing to note the dynamic development of telecommunications, information technology and the audiovisual media, and quite another to credit them with changes in culture and in other areas of social life. Technical innovation is a by-product - only rarely a driving force - of transformations whose origin lies elsewhere. Technology provides such changes with new areas of opportunity, encouraging activities which would otherwise have been blocked by the dominant organizations in the sector and weakening the barriers set up to defend vested interests or to safeguard specific practices. It is important to look beyond the usual thought patterns, which see the issue as one of adaptation to technological changes, and turn our attention instead to changing trends in cultural production.

ICTs may strengthen cultural production and help it to gain widespread recognition, or conversely, they may cause it to be confined to a restricted area or even to decline. Throughout the world, cultural production tends to be community based. Anthropologists, historians and cultural sociologists rightly emphasize this recurring feature which explains the persistence of locally-rooted and even non-commercial forms of culture, as well as the different ways in which viewers interpret television series, according to their own culture. In other words, despite more than two centuries of industrialized cultural production and despite the aggressive strategies of the increasingly-powerful communications groups, independent forms of entertainment still thrive all over the world. Folk productions are not declining - far from it: books and newspapers are published, films and television series are produced and music is recorded independently of the major companies.

The pessimistic forecasts of the imminent globalization of cultural production heard over the last two decades have been proved wrong - or at least exaggerated. Culture remains an area of social activity in which industrialization meets with resistance and constraints, even though increasingly governed by markets, and despite a growing trend towards the internationalization of production and the transnation-alization of subjects, styles and standards. The fact that predictions have not so far been fulfilled is, however, no guarantee for the future. ICTs and the momentum they undoubtedly generate will eventually provide the means to move decisively ahead towards the industrialization and internationalization of cultural production. What in the past was only partly successful is now allegedly resulting directly from ICTs. In other words, what the mass audiovisual media merely launched in the past has now become firmly established and universal through ICTs, which have proven to be more successful at sidestepping the cultural structures of countries and peoples, and certainly more favourable to the worldwide circulation of transnational flows. Individual tools of communi-cation and/or access to data and ready-made programmes are ushering in sweeping transformations in cultural production.

The history of the cultural industries is by no means linear. Until recently, it has progressed unevenly, whenever production techniques were adopted and made widely available, so that works of literature became accessible as books, musical works as records, and films and audiovisual works as screenings in cinemas or as video cassettes. The industrialization of culture has, however, encountered difficulties and setbacks. For producers and publishers, the values attached to culture usually act as 'restrictions', or at any rate as contingencies, for which marketing techniques have yet to be found. It has been more convenient to market the equipment used to deliver cultural works than to introduce successfully on to the market works available either in a physical medium or accessible through projection or broadcasting by the audiovisual media. So the picture of the relationship between culture and industry today presents sharp contrasts. There are striking divergences between:

— developed and developing countries;
— countries which give more or less free rein to market forces, and those in which the public authorities provide impetus and co-ordination;
— countries where the market and industry are the major controlling factors in the cultural sphere, and those where small-scale productions, live entertainment with a popular appeal and non-commercial productions predominate;
— the various art forms: it is very difficult to create added value in dance, poetry and narrative performances, as opposed to filmed enter-tainment, which should be considered an activity straddling art and industry;

— individual access to works and live entertainment;
— cultural productions broadcast through the mass audiovisual media and those which, once published, are bought by consumers.

ICTs are capable of strongly reinforcing trends begun at least two decades ago, thereby giving a decisive boost to the expansion of cultural industries. The areas in which ICTs can make an impact need to be carefully evaluated, however. Contrary to what professionals often claim, it would be wrong to expect multimedia production to create innovative products - apart from games -and important markets in the short term. Innovative products of a truly multimedia type are still too rare and production costs remain too high for rapid progress to be expected. If there is not a rapid change in product design and production methods during the next decade, ICTs will probably not be instrumental in extending the scope of commercial and even industrialized culture, since with the products which are currently available they are increasingly marginalizing certain cultures, standardizing universally accessible products and excluding those who lack the purchasing power to be ICT consumers. The solution to this problem would appear relatively easy and involves speeding up product flow, which is the second major feature of the development of the cultural industries.

Distance-shopping for products hitherto almost e xclusively supplied by specialized end-product distributors or by mass distribution outlets, and downloading products accessible from a distance are two activities that have been made possible by the efficiency of communications networks and the growing sophistication of access terminals. The scope of this new electronic commerce is still difficult to gauge, but there is no doubt that in 1997-98 its growth was such that the major distribution groups all rushed headlong in that direction, preceded by small pioneering firms. This strategic choice was made despite the fact that monetary transactions on the Internet are still not totally safe and that arrangements for remunerating artists and publishers are still under negotiation. Most specialists consider that cultural products are particularly well-suited to this new form of distribution. There are substantial advantages for consumers: real-time access; the opportunity to browse through all available catalogues and products; access to catalogues of foreign firms which used to be accessible only, if at all, with long delays involved. In other words, these new ways of accessing cultural products are of special interest both to the consumers least concerned by mass products, and to specialist publishers and producers.

Paradoxically, the people who are apparently the most critical of this rapid change in the distribution of cultural products ('cultivated' members of society, let us say, who prefer the existing methods of accessing art forms, and also scientists) are those who, for the moment, find it most advantageous to purchase directly via the networks. There is an additional paradoxical aspect to these immediate advantages for the most selective and demanding users. Producers and artists who offer the most distinctive products and who frequently encounter great difficulty in distributing the works they create and publish will obviously be tempted to bypass the large broadcasters and distributors in order to contact consumers directly. This is all the more likely to happen because, throughout the history of the cultural industries, artists and their publishers have tried to eliminate intermediaries with little success. They themselves then joined that enviable group, thus only reinforcing the existing oligopolies.

The current trend consists mainly of users purchasing products after browsing through catalogues on publishers' sites and on the sites of artists acting as producers of their own works. Soon it will be possible to download actual works, meaning that publishers of books and recorded music will no longer need to reproduce and record the work in a physical medium. This is not a utopian prospect; it is already possible to access literary and musical works in this way, or even extracts from them. This phenomenon raises two essential questions about cultural products. Is a work on offer a finished product just because it has been stabilized in a medium? Next, have the forms of monetary payment by consumers, here encouraged to buy works piecemeal, and 'paying-as-they-go', been fully considered? These are not merely technical innovations which, by encouraging a commercial shift, coincide with changes in systems of accessing works. Here technical innovations are changing the very substance of cultural products in various ways, especially by calling into question their finished appearance, an aspect considered fundamental up until now.

In modern societies, the values and images associated with the concepts of leisure, information and culture are still sharply differentiated. Although these concepts are evolving and now sometimes overlap, it would nonetheless hardly be possible to confuse them taken in a strict sense. They have relatively stable definitions: infor-mation is more ephemeral than culture, and this idea orients the functioning of those who produce news and information; leisure time is primarily for entertainment, but leisure pursuits encouraging the development of cultural activities are judged as positive. Although the trend began earlier, ICTs are causing these distinctions to grow more and more blurred, and in some cases it is becoming difficult to distinguish between information products and cultural products. Methods of creation (by journalists or artists, with an increasing use of streamlined methods), production structures (more and more often controlled by the same multimedia groups), distribution procedures (via networks or in the form of products accessible on- or off-line), consumption time and conditions of consumer use (use is often individual and involves interactivity with the hardware acting as intermediary in programme communication) are now closely related, not to say identical. These similarities cannot be considered as either secondary or insignificant.

Behind all the confusions and convergences, the question that arises is more than ever that of mass culture, although in forms different from those observed in the 1960s when the phenomenon became evident. As the transmission of standardized and increasingly transnational mass culture boosted by ICTs develops, cultural distinc-tiveness tends to be maintained within an elitist, or at any rate a niche, context. Although distinctive cultural production is not declining, and examples of it which successfully express popular and national identities continue to receive considerable attention, they enter into competition with mass products that are easily distributed on communication networks.

The struggle is to some extent unequal, not because of inadequacy of supply but because the major communications groups are strengthening their hold on the circulation of products. Insofar as future developments can be inferred from the situation today, culture seems to be going the same way as leisure and information. For most people, ICTs have not yet become part of a familiar landscape. They have already led, however, to some fairly radical changes in the functioning and regulation of public services, resulting in considerable hesitation on the part of those responsible for cultural policy. An increase in the supply of commercial products almost inevitably leads to the reduction, if not the disappearance, of non-commercial products (heavily subsidized by public funds) and of semi-commercial products (e.g. small-scale productions and those enjoying tax and customs protection). A movement of this type has recently become apparent in television production. In response to competition from powerful new commercial channels and in order to increase their output of programmes to compete with the commercial channels, with approximately the same resources, the managers of public channels have been obliged to subcontract production, placing it in the hands of new private producers for whom the origin of the orders they handle is of no importance. As a result, production variety in public television programmes has declined sharply, and some types of programme are less available or have actually disappeared since they are not commercially viable in a situation where audience ratings are all-important. Poetry and other literary programmes have disap-peared or been banished from prime-time slots. Even once-popular variety shows have been downsized or incorporated into shows with a wider appeal, since the popularity of a group or singer is no longer sufficient in itself to attract a mass audience. Over the past ten or fifteen years, some national cultural industries have continued to function and have even grown as a result of certain trends in public policy.

In the European Union, community-wide or domestic measures have undoubtedly been beneficial to certain industries in Member States, such as book publishing (the French system of retail price maintenance for books has allowed a network of specialized bookshops to survive), recorded music publishing (radio stations are obliged to play a quota of nationally-produced songs and music), cinema and audiovisual production (a system incorporating various forms of aid to production) and television production. There is, however, no guarantee that these hitherto undeniably advantageous effects will continue in the future. Not only do quota systems run into operational difficulties attracting regular criticism because of their singularity in world commercial transactions, but the fact that programmes can now circulate on the networks and largely escape control from public authorities will tend to reduce the effectiveness of support measures for small- and medium-sized industries operating mainly within a specific cultural or linguistic framework. This does not mean that such measures are unnecessary, but that their effectiveness may become more uncertain, simply because con-sumers will tend to diversify demand, including demand for products from the sphere of high culture.

A number of authors representing different schools of thought have long been predicting a trend towards the individualization of social behaviour, and the emergence of this trend does seem to be reinforced by ICTs, particularly in the developed countries.

ICTs, unlike the media that developed earlier, usually require user-consumers to take active steps not only to access a specific programme but also to interact with it in order to obtain responses tailored to their personal requirements. To this extent, ICTs, more than any other earlier technology, promote a one-to-one correspondence between the content on offer and the user's demands, which are apparently unlike those of anyone else. Where ICTs are involved, cultural consumption is not simply a question of narcissistic contact with works. Likewise, collective forms of consumption are not doomed to disappear, because individualization does not necessarily spell a decline in public performances and other forms of collective entertainment.

Despite predictions to the contrary, the proliferation of individual delivery systems for recorded music has not led to a decline in concerts of all types of music. Complex interactions and processes of mutual i n fluence are observably taking place between different types of music, but it is not to be expected that this will have any practical outcome, although new linkages can be foreseen. ICTs encourage a fragmented, episodic pace of consumption which has little in common with the heyday of collective cultural consumption and this tendency may lead to a number of fundamental cultural changes. One consequence may be for increased cultural consumption to go hand in hand with a decline in discussion and debate focused on cultural productions. It is also possible that performances may become almost exclusively opportunities for the 'promotion' of cultural merchandise, something that would be inevitable if public aid earmarked for live performances were reduced or even abolished, with those responsible taking as their pretext the variety and quantity of commercial products available. This would mark the culmination of the process by which culture becomes a commercial product and, as noted above, is absorbed into the leisure and information sectors.

ICTs have promted American English to the status of an international language. According to a 1998 Euromarketing study, more than one of every two Internet surfers (58%) uses English. Other languages - Spanish (8.7%), German (8.6%), Japanese (7.9%) and French (3.7%) - are much less commonly used. These data, which are obviously difficult to establish, have nevertheless been confirmed by other studies. However, they are difficult to assess because change is so rapid in both hardware and habits. Furthermore, it is not possible, on the basis of data on the preferred language of communication (used for electronic mail and forums), to draw similar conclusions regarding trade in cultural products. For example, after an initial phase when English.. predominated, the satellite channels are now beginning to diversify into other languages. Will automatic translation systems contain applications that will help to reduce misunder-standings between languages and promote genuine multilingualism?

In the past, research has proved disappointing, and researchers are at present cautious, pinning their hopes on the development of computer-assisted translation systems. This indicates recognition of the fact that, in this field, approximations and simplifications are not sufficient, because they undermine languages as vehicles of specific cultural features. Only strong support for translation from producers and from authorities responsible for public policy can help to bring about a more balanced flow of cultural works. If this is not forthcoming, English and probably a few more of the most widely spoken languages will eventually dominate the cultural marketplace. Support for translation and for the development of translation aid systems will not, however, be enough in the absence of systematic efforts to support productions suitable for distribution outside their country or region of origin, and support in distributing them. Recent experience has shown that the successful exportation of films, works for television and sound broadcasting or even literary works, depends as much on international standards being met as on the availability of an appropriate distribution system. It is also important for production costs on national broadcasting networks to have been absorbed in advance, which explains the recent success of Brazilian and Mexican series outside the areas where they were produced. The same factors are likely to be important in the future.

Although in theory the new communications networks guarantee wider distribution, and even distribution that crosses borders and cuts out intermediaries, which may be helpful for niche products that may even be boycotted in their place of origin, it would be illusory to think that world markets will develop on this basis. Direct transactions will leave room for a few new actors, but free flow will probably lead to a rapid jump in production costs and a concentration of distributors. The major communication groups are preparing for this eventuality. Their uncertainties about the future focus on the plans of their direct competitors rather than on the emergence of newcomers or the development of direct commerce that would be beyond their control.

ICTs undeniably complicate a number of issues facing the production of culture in all its diversity. However, the more or less rapid spread of hardware and networks is not the only factor involved in sustaining artistic pluralism within countries and regions and keeping a variety of production methods in place. Blithe and often exaggerated statements to the effect that we are now living in an information society which will impose its ground rules on culture should be viewed sceptically.

For complex reasons that research into technical innovation has already begun to elucidate, technical developments seldom fulfil expert predictions. In other words, ICTs do not contain a blueprint for the future. They will not be the sole factors shaping the future. If, however, those responsible for cultural production worldwide adopt a wait-and-see attitude and fail to grasp the importance of what is at stake, then culture will be increasingly subservient to the norms of mediatized communication and information flows.

Public cultural policies are important and should be reinstated in areas where the field has been left free for the commercial audiovisual media and transborder flows of information. Public action will have little effect, however, if it is confined to protective and defensive measures which are in any case liable to be inefficient, if not impossible, to implement. This issue is likely to give rise to considerable debate and to social movements and even 'resistance' instigated by artists, producers, users and political leaders.

During the 1993 General Agreement on Tariffs and Trade (GATT) negotiations, France strongly supported the idea of 'making an exception of culture' and provisionally won its case, arguing that with the opening of world markets, cultural products should not be treated like other commercial commodities. Since then, the question has been asked worldwide, some counting on a de facto abolition of the 'cultural exception' by technological development, while others are waiting for international organizations and regional bodies to come out in support of cultural pluralism. Another crucial issue being debated is the future of copyright and authors' continuing control over their works in an environment complicated by the emergence of multimedia products. Many developing countries have not yet stated that they are directly concerned by these discussions, which nevertheless do affect them. In order to maintain cultural pluralism, it is pointless to avoid or side-step negotiations on these basic issues. An attempt was made to do this in the case of the Multilateral Agreement on Investment (MAI) by seeking to negotiate behind closed doors within the OECD. What are needed are balanced agreements that acknowledge the distinctive nature of cultural products and do not treat them simply as industrial goods.

## Impact of ICTS on Mass Media

The increasingly interconnected nature of the mass media industry has encouraged integration in a complex variety of reciprocal connections. The first set of logical mergers were between different types of media, such as television with radio and the press to create multimedia conglomerates. Mergers of media entities across countries and regions also occurred in order to attain the critical business strength needed to reach global markets. A few hardware manufacturers made strategic moves to purchase media houses producing films, music and other media products. The strategy has paid off in recent years through co-ordinated releases of new generations of hardware, such as Digital Versatile Disk (DVD) players, digital television receivers, and digital music recording devices, together with media products and software to be played back on these new devices. The synergy is mutually beneficial. There would have been no demand for the hardware had there been a lack of attractive media products to play on the devices. Without the devices, the media companies would not have been able to launch a new generation of both old and new releases which help to produce a new stream of revenue for the existing libraries of intellectual properties owned by these media companies. In an interesting new twist to integration and convergence, a major television sports broadcaster has mounted a takeover of one of the most famous soccer teams in the world.

The business strategy is to acquire full control of broadcast and merchandising rights through the outright purchase of the team. If this proves to be a profitable move, it is not impossible that media companies may begin to purchase properties as diverse as universities and restaurants in order to secure publication copyright for scholarly titles and courses, and broadcast licences for cookery programmes. There are two sides to integration and convergence. The case against concentration of ownership can be seen within the printing sector in the United States, where extensive research shows that the twelve largest newspaper conglomerates control almost 50% of all circulation in the largest market for news in the world. In book publishing, nearly every one of the major commercial publishing companies has also been bought and assimilated into larger businesses, while independent bookstores are being forced out of business by a handful of powerful national chains. On the positive side, the sale of some cable television companies, again in the United States, helped raise much-needed financial investment to upgrade distribution systems. The merger of Cable News Network (CNN) with the Time Warner group led to the pooling of extensive complementary intellectual properties, particularly in politics, which has resulted in service on the Internet which offers more depth and data than either company alone could have provided in the print and television media.

The expansionary trends of the media conglomerates, coupled with media technologies which ignore political borders, have forced government-owned media organizations in newly developed and developing countries to respond. Within Asia, the response has generally been to privatize state-owned media in an effort to make them more competitive and attractive to domestic audiences. However, the privatization has been structured in such a way as to protect, preserve and empower traditional cultural values and maintain key government controls at the same time. The Association of South-East Asian Nations (ASEAN) plans to launch a regional direct satellite television broadcast channel, but these plans have been delayed by the financial problems affecting the region. When several large media organizations merge, they create media giants with huge audiences. This is an invaluable asset which helps in maximizing advertising revenue, the lifeblood of media. The merger also pools strategic market information, such as subscriber and advertiser databases which are used to cross-sell media products and advertising across the organizations within the merged group.

Direct selling is probably the showcase innovation of the private sector on the Internet, and Amazon.com is the current undisputed star in this area. Amazon began by selling books over its Internet site and subsequently added music CDs to its virtual bookstore. A strictly e-commerce company, it has captured the imagination of media organizations because it has so far dealt exclusively with media products. Indications are that it will diversify rapidly to include a large variety of consumer products. Every media company with tangible products must at some time have reviewed Amazon's methods for possible emulation. The arithmetic of direct selling to consumers worldwide without having to incur high distribution and sales discounts is an irresistible idea, and an increasing number of media organizations with such product offerings have launched websites to test the feasibility of an Amazon-like sales outlet. Attracting less public attention, but perhaps more exciting in its technological concept, are the direct sales facilities of newspapers, journals, bulletins, and electronic book publishers. Amazon uses the Internet only to promote and process payment for a product before resorting to fairly conventional means for delivering a traditional tangible media product to the consumer. In contrast to this, specialized publishers are offering articles and information from databases for instantaneous downloading over the Internet immediately after a payment has been made at the site. Book publishers will be interested in the mass consumer response to several electronic book devices introduced recently into the market and which approximate the size and feel of a book. They lend themselves to the direct sales of books bought and delivered over e-commerce systems.

The North American book distributor Barnes & Noble has negotiated with various publishers to offer more than 1,000 titles of books for downloading into the devices which have a capacity to store anything from about 1,000 to 500,000 pages at any one time. The e-book is an extremely attractive proposition to book publishers because they do not have to commit expensive print-runs and warehousing expenses up front; the risks are therefore minimum. This may encourage publishers to accept a larger number of manuscripts for publication. It is likely to facilitate the release of traditionally short-run titles, which have been rejected in the past because they were not viable as expensive paper editions. The e-book is a viable alternative to currently available publish-on-demand technology, which deploys high-speed photocopying machines linked to collating and binding machinery. The music industry may be on the verge of making similar direct sales. A new device which downloads and stores audio digital recordings from the Internet for repeated playback by consumers was cleared for public sale, following a court hearing in California (USA) to determine if such a device threatened the preservation of intellectual property rights. The legal endorsement for the device has set the stage for music to be sold directly to consumers, a song or 'cut' at a time, rather than as a compilation on longer-playing discs.

The ICT technologies, with their capacities to move multimedia content anywhere in the world in a matter of seconds, provide the perfect tool for gathering and disseminating news. They have become available at a time when the demand for news is rising rapidly, fuelled by the globalization of trade and national economies. The increasing demand is not only for a greater quantity of news, but also for higher speeds in reporting. This in turn presents developers of technology with demands to increase further the capabilities and speed of technologies creating a powerful iterative spiral, which will lead to even more, exciting technologies.

The demand for news worldwide is obviously not uniform. Television ratings are a good yardstick by which to gauge demand. News dominated the top ten television shows in the United States; soap operas of different origins seem to be the most popular in many other countries such as Brazil, China and South Africa. The demand for news over the Internet appears to be more consistent, with nearly all news sites registering the most number of visitors and Web pages downloaded. Sites operated by newspapers around the world commonly serve millions of page views per month. While most newspapers continue to provide only text and static images on their Web sites, radio, television and other integrated media conglomerates now offer multimedia content which includes text, and sound and video clips of headline news items. Several financial publications operate Web sites with access to constantly updated databases and expert systems software to work out investment computations and projections.The exponential growth of the Internet as a channel for distributing news can be seen in the meteoric growth of Microsoft National Broadcasting Company (MSNBC), a joint venture between Microsoft and the National Broadcasting Company (NBC) in the United States.

By the first quarter of 1998, the news site served an average of 300,000 users a day with the occasional peak of 900,000 users. Together, they read 200 million pages a month. Measured by reach, it is the fifth largest daily newspaper in the United States. The top four Internet news sites in the country, MSNBC, USA Today, CNN and the American Broadcasting Company, collectively serve up 700 million pages of news every month.Direct satellite broadcast television news channels and the Internet benefit people around the world who want and need to have constantly updated access to news at all times of the day and night. This has caused an insatiable demand for the latest news, which has in turn created the twenty-four-hour news cycle. News operations in the past were governed by deadlines fixed by printing schedules, in the case of newspapers and magazines, and airtimes for conventional radio and television news. Now that the audience is global, every moment of the day and night is the deadline.

All news programming is by definition perpetually incomplete and in a state of being constantly updated. The concept of 'the edition' has been replaced by devices such as 'breaking news' when regular programming is interrupted in television broadcasts, to show raw, unedited, live coverage about a still-evolving news item. On the Internet, a similar device is the 'latest' button, which provides a link to a usually brief write-up of a news event in the making.

The race to be the first on the air or the Internet with the latest news has created an approach to news reporting which concentrates on the immediate, the gathering of eye-witness accounts, and a follow-up of the event as it unfurls and draws to a conclusion. Attention to analytical writing is on the decline. The twenty-four-hour news cycle, and the quickly changing news agenda discourages reflection, research and analysis which take up time, and time is what the editorial departments no longer have to spare. The result is sometimes news reporting which is disconnected from history and the broader geo-political context.

Merrill Brown, editor-in-chief of MSNBC, is mindful of the shortcomings of journalism on the Internet: 'Those of us producing for the Internet haven't always been as thoughtful as we might about putting today's incremental developments into perspective'. Andrew Heyward, the News President of Columbia Broadcasting System (CBS), one of the dominant North American television networks, is equally conscious of television's short-falls. He identified 'the seven daily sins of television news' as imitation, predictability, artificiality, laziness, oversimplification, hype and cynicism. Although not as profitable as the owners would like, the media is now big business. During the con-centration of ownership, large financial investments were made; owners are now eager to recover them. At the same time, profits at many media organizations have shrunk. Together, these considerations have driven media organizations to cut costs and this has harmed the quality and diversity of news services. Among the elements which have been slashed are foreign news bureaus and the amount of original reporting undertaken.

News editors are also less inclined to dispatch their own reporters, photographers and camera crews to cover an event, preferring instead to rewrite wire service stories and tap into news footages from large television news agencies. The impact of this trend is a significantly reduced plurality of perspectives, especially where international news items are concerned. The same television news clips are broadcast by stations around the world, while the same wire service dispatches are carried in the newspapers. This provides the handful of international news agencies which feed the media.

organizations around the world with an oligarchic grip on the flow of news and views. This phenomenon is called 'consonance'. Research has shown that the further the geographic origin of a news story from the editorial headquarters of a media organization, the higher the level of consonance. Accuracy of information has also suffered as a result of cost-cutting. Reduced resources to research and confirm facts often lead to the release of not completely accurate news and content. Because of the concentration of media organizations, the effects of these errors are sometimes multiplied across several types of media.

The internet has helped to inject a level of participation in news commentary that did not exist in the past. This has taken the form of on-line polls of audiences and readership using customized software which operate at media Web sites. Although the samples reached in such polls are obviously not always representative, they are nevertheless a good beginning in getting the audience involved in developing opinions about issues. Current methods of polling offer audiences a limited number of choices, identified by the media, which may not always reflect the full spectrum of opinion. Perhaps more useful are facilities at certain Web sites for visitors to post their comments.

The Internet version of a phone-in comment is tremendously popular with interactive radio stations. A few international radio and television broadcasters have also started to use audience postings of comments at Web sites and e-mail, as a channel for the audience to participate in live panel discussions. The low cost of using the Internet by the audience permits this form of live participation at the global level on a sustained basis and, if facilitated effectively, will contribute towards enriching the diversity of perspectives in all types of media. In participation of a different sort, newsmakers now use the Internet to reach out directly to the audience, bypassing the media. Court judgements have been posted directly on to the Internet, as have copious copies of original reports of powerful government-appointed investigating authorities. Such events are being multiplied in a myriad of other events, ranging from the birth of a child to an expedition on Mount Everest.

Quick-thinking editors and their lawyers have found an interesting innovative use of the new technologies in the defence of the freedom of the media. United States media organizations subpoenaed for information in their possession, and faced with prospects of losing the legal challenge when no important information is involved, have been known to publish photographs or broadcast footages so that they can honestly claim that they have not surrendered any unpublished material to the prosecutors. The new technologies allow news stories and pictures to be rushed out into the public domain over the Internet. Television companies, faced with the need to place uninteresting and insignificant footage in the public domain, have been known to broadcast this material in the middle of the night using an idle satellite

## ICTs and New Business Models

The evolution of a new generation of media technologies has presented the industry with many possibilities for developing new products and services. One of the key factors, which will help in identifying the final list of possibilities for sustained development, is the parallel evolution of business models applicable to these possibilities. The successful candidates must operate with a business model acceptable to the consumers, and attractive enough to engage media operators. The private sector has long been the primary source for funding of media operations in North America, Europe, Latin America and parts of Asia. With the trend towards the privatization of state-owned media in developing countries, the importance of feasible business models has become critical in the determination of how technologies will be deployed, and what content and software will be offered.

### Large Audience-base Models

This type of business model dominate for traditional mass media such as radio, print and television. It has also become one of the most important models for Internet-related operations where 'portal' sites offering search and directory services to the vast World Wide Web have drawn the largest number of users and attracted the keenest interest of investors and entrepreneurs. As with most broadcast organizations, portal sites do not charge for their services but instead derive revenues by selling advertising. In this model, the number of people served by a particular Web site or broadcasting station is crucial in determining the pricing of its advertisements, and eventually its revenue. The same applies to the print media products, which are often priced for the recovery of production and distribution costs only, while profits and creative costs derive from the sale of advertising space. These numbers will take on added importance when members of the audience become customers and shoppers on interactive television and electronic commerce Web sites.

New audience-tracking technologies and sophisticated audience research methodologies enable these numbers to be desegregated into groups with a high propensity to purchase certain goods and services. These advances are a boon to advertisers and will pose a challenge to media operators, who, in the past, were required simply to broadcast advertisements for large groups. The tougher requirements are already apparent in advertising at Web sites where operators are required to broadcast advertisements in the style of banners at the top or bottom of specified Web pages, and the cost of advertising was determined by the number of people accessing the page and seeing the advertisements. This simple formula became a little more complicated when the cost was computed on the basis of the number of people who clicked on the banners and visited the Web site linked to the advertisements. The third formula which is evolving will very likely be based on a sales commission on the actual business transacted as a result of exposure, click-throughs and other services rendered at a particular Web site or page.

### E-commerce Model

This model has captured the imagination of the industry as the most appropriate for the Internet with Amazon.com as the premier prototype company. Amazon has demonstrated both the potential and the problems of this model. Although it is popularly cited as the most successful electronic commerce company, it has not shown a profit since it started. On the contrary, Amazon consistently incurred losses until the end of 1998. Researchers who have tracked the company's exponential growth blame the losses on high operating costs. Its economics also reveal a staggering scale of business. One study estimated that Amazon must generate $1 billion in annual sales just to break even. The electronic publishing model has been given a boost with the introduction of secure electronic transaction software and systems which permit payments for purchases made over the Internet to be securely debited from credit card accounts.

### Subscription Access Model

Subscription access was widely thought to be a promising model for the electronic publishing of newspapers, magazines, journals, and newsletters.

The typical electronic subscription begins when a person makes a lump sum payment to the publisher, who then provides the subscriber with a unique password, which then permits access to each new edition at a Web site. Although the model was based on very sound logic, it was not well received by people on the Internet. For the present, it appears that the founding principle of the Internet for the free sharing of information still persists. Major content providers have indefinitely postponed plans to charge for access to their sites. To replace this, some providers require visitors to their Web sites to register themselves, or encourage them to click on the advertisement of a corporate sponsor, as a kind of surrogate payment. This has caused extensive revisions in the business plans of media companies started for the exclusive purpose of publishing and selling information over the Internet.

Conventional media, such as television news networks, newspapers, and magazines, which extended their dissemination by recycling selections from their products on the Internet, have enjoyed more prosperity. Their sites regularly clock up millions of visitors each month who in turn each read several Web pages and help to generate a total of many millions more 'page views' for the sites. While people visiting the sites do not pay for their visits, the media organizations running the sites benefit in several ways: firstly, by promoting or cross-selling their flagship on printed or broadcast products via free access to excerpts at the Web site; secondly, by offering archival information such as previously published articles or transcripts of old broadcasts for sale through an automated system of database searches and electronic commerce; and lastly, by exposing people to this new outlet for their products and developing a global audience for future sales of a new generation of products and services.

The Economist and The Wall Street Journal now publish simultaneously on paper and on the Internet. There are many sites on the Internet that provide news. One type are the special Internet news sites, often provided by the Internet Service Providers (ISPs). Others are sites maintained by already-existing newspapers as a complement to the printed product. One can find daily newspapers and news-oriented periodicals from almost every country in the world, representing a mixture of national and regional/local press, as well as news targeting specific groups in terms of language and/or ethnic belonging. One significant result of this development is that, whereas the possibilities of access to traditionally disseminated press were limited, especially concerning regional and local press, news via the Internet means worldwide dissemination.

The former provides subscribers to the printed product and free access to the Internet edition. The latter sells subscription access to both versions. Some newspapers have used their Internet editions to enhance the attractiveness of their classified advertisements by offering package deals, which place such advertisements simultaneously on both printed and Internet editions. As a result of the financial crisis of 1997/98, media organizations throughout Asia, Latin America and Russia are struggling to stay solvent. The media in these regions have been hit by high increases in repayments of bank loans and other forms of business financing, increases in the price of newsprint, and significant drops in advertising revenue, as companies have trimmed their advertising budgets as part of cost-cutting measures. Internet services providers have suffered from the crisis as well, because the values of local currencies have fallen in relation to the US dollar in which the cost of telecommunications connections are calculated. At the same time and for the same reason, the cost of imported networking hardware and soft-ware has also increased.

## Challenges of ICTs

The reality of the widening gap in digital capacity raises the serious concern that the poorer countries may not be able to overcome the financial and technical obstacles now limiting their access to the digital technologies. In early 1995 the concern about the ICT-gap inspired many public and private donor institutions to propose plans for the elimination of digital disparity. For example, the World Bank established the Information for Development Program to assist developing countries with their integration into the global information economy. In the same year, the ITU established World Tel, an ambitious project to generate private investments for bridging the telecommunications gap in the world by developing basic infrastructures. WorldTel aims to establish 40 million telephone connections in developing countries over the next ten years, which will require a minimal investment of $1,000 million.

The equitable sharing of communication infras-tructures (the electronic highways systems created by carriers such as satellites, cables, fixed lines and mobile transmissions), computing capacity (computers, peripherals, networks), information resources (databases, libraries), and ICT-literacy (intellectual and social capabilities to deploy ICT in beneficial ways) will require an enormous effort from the international community. Massive investments are required for the renovation, upgrading and expansion of networks in developing countries, for programmes to transfer knowledge and for ICT skill training, in particular for women. In 1985, the Maitland Commission estimated that an annual investment of $12,000 million would be needed to achieve the goal of simple, universal access to a telephone early in the 21st century. To attract private funding, countries will have to liberalize their ICT markets and adopt measures in favour of competition. In this context the World Bank has recommended the creation of investor-friendly business environments, the protection of investments and security for repatriating revenues. When public companies are privatized there may be losses in revenues as a result of the change in service charges or severance payments. The World Bank is therefore proposing to finance the adjustment costs of the adoption of liberalization schemes in individual countries.

The World Bank's policies are characterized by a strong emphasis on economic growth and a key role for the private sector. The expectation is that, in a sufficiently free market, economic growth will also benefit the poorer sector of society. In this context the contribution of the ICTs is to provide the essential infrastructure for economic development. This position bypasses the question raised earlier, as to whether the deployment of ICTs does indeed lead to growth in economic productivity and, if so, whether such growth will be equitably distributed. The governance of ICTs is in fact left to freely operating private entrepreneurs. The basic assumption is that a country's telecommunication infrastructure can be managed by private companies and that, whenever parts of the network are unprofitable, the state will provide the public means to ensure that no citizen is disenfranchised.

There is some debate about the expectation that private funding will create worldwide equity in the access to and use of ICT resources. It appears in any case that the international community and national governments of affluent countries need to bear in mind that problems stem not only from a lack of financial resources but also from a lack of political will. Creating adequate access to ICT resources worldwide should not be a problem in a world economy with income amounting to roughly $22 trillion. The core issue is that expenditures for development assistance represent only $55,000 million and thus a mere 0.25% of this income. The annual costs for all developing countries for adding one thousand million telephone lines, subsidizing over 600 million households that cannot afford basic telephone charges, providing PCs and access to the Internet for schools over a period of ten years could represent from $80,000 to $100,000 million. This is not a prohibitive level of funding. It represents about 11% of the world's annual military expenditure, about 22% of total annual spending on narcotic drugs, and is comparable with the annual expenditure on alcoholic drinks in Europe alone. For a variety of political and economic reasons, many donor governments are presently reducing budgets for financing ICT-development.

Between 1990 and 1995, multilateral lending for telecommunications decreased from $1,253 million to $967 million. Bilateral aid for telecommunications decreased from $1,259 million in 1990 to $800 million in 1995. Financial obstacles are, however, not the only concern. The transfer of ICTs also raises questions about their appropriateness and about the capacity of the recipient countries to make the best use of them. Over the past few decades the prevailing international policies on technology transfer have placed formidable obstacles in the process of reducing North-South technology gaps; the present discussion on the ICT-gap provides no convincing evidence that the technology owners will change their attitudes and policies towards the international transfer of technology.

There is no indication that current restrictive business practices, constraints on the ownership of knowledge, and rules on intellectual property rights that are adverse to developing country interests are radically changing, and there are no realistic prospects that the relations between ICT-rich and ICT-poor countries will change in the near future. The question must be raised as to whether there can be any serious reduction of the ICT disparity, given the realities of the present international economic order. It may well be an illusion to think that the ICT-poor countries could catch up or keep pace with progress among countries in the Northern hemisphere, where the rate of technological development is very high and is supported by considerable resources. This is not to say that poor countries should not try to upgrade their ICT-systems. They should, however, not act with the unrealistic expectation that those who are ahead are planning to wait for them.

In most countries, the problems concerning access to ICTs are handled through public policies based on an already defined technological environment. Developing countries thus find it difficult to assess what digital technologies would be appropriate for their specific development strategies.

A problem compounding this situation is that in many cases 'peripheral states seem to have no disinterested non-governmental organizations to advise them on telecommunication technology and the social objectives of regulation that would safeguard those interests that private profit will not meet. Without adequate regulatory intervention to ensure accountability to the general public, market forces that respond to those groups with purchasing power are bound to generate unequal development'. Most developing countries lack the capacity to identify appropriate digital technologies, and, to make matters worse, there is also a critical absence of co-ordination of 'digital' policies among the developing countries themselves.

It is essential to recognize that planning for the adoption and deployment of digital technologies can no longer be a local affair. Global negotiations, such as the recent General Agreement on Tariffs and Trade (GATT) Uruguay Round of Multilateral Trade Negotiations, heavily affect national technology plans, while the processes of globalization. Information and communication technologies and social processes determine the playing field for local actors. As a result, local planning has to take into account the effects of global forces, possible only when planners in the periphery pool resources and mobilize a constituency that counteracts the Northern dominance in global planning. Since today's globalization process is largely determined by Northern forces, 'many developing countries do not obtain a fair share of the benefits of globalization, and some actually suffer net losses'. The North is in control not only due to strength but also because of lack of co-ordination in the South. National technology policies are largely determined by the work of global institutions and their rules and standards. It is vital that developing countries participate more forcefully and effectively in these institutions.

## Future of the ICTs

The future of the ICT sector is exciting. These are unchartered waters open to creativity, innovation and entirely new ways of working, interacting and learning that should appeal to women and men alike. The Institute for the Future identifies six drivers most likely to shape the future workforce: longer life spans; a rise in smart devices and systems; advances in computational systems such as sensors and processing power; new multimedia technology; the continuing evolution of social media; and a globally connected world. The ICT sector clearly underpins this future.

The ICT sector remains a buoyant and growing sector for employment and a key sector underpinning both national and international development. Employment in the ICT sector has continued to grow significantly in recent years. This growth, however, has not led to a parallel increase in women's presence in the ICT labour market, with the male-female gap being particularly pronounced at senior levels. In comparison to the general growth of the sector, women's employment figures in advanced economies are actually in decline, which suggests that the issue is not just an entry level problem but also one of demotivation, of retention and/or promotion of women within the sector at many levels.

The perception in most countries that the ICT sector is a male-dominated industry still persists. Males dominate most high-value and income jobs in the ICT sector. Research conducted for this study in both developed and developing countries found classic cases of vertical gender segregation, with women more strongly represented in lower level ICT occupations than in higher status and higher paid arenas. Although women are making inroads into technical and senior professions, there remains a 'feminization' of lower level jobs. On average, this research found that women accounted for 30 per cent of IT operations technicians, a mere 15 per cent of ICT managers and only 11 per cent of IT strategy and planning professionals.

Human talent with the right skill sets will continue to be the key for the building of a vibrant and diversified ICT sector. That talent pool will need to be enriched by the building and training of non-discriminatory human capital primarily in universities, research and development centers and trade or 'applied' schools to respond to the evolving ICT industry.

Because the ICT sector is fast-paced and dynamic, the skills of the ICT workforce need also to keep up with the pace of change. This suggests that ICT qualifications need to be extended to include a much broader spectrum – which in turn suggests that there may potentially be more employment openings that might attract the attention and interest of girls and women.

The ICT sector needs to invest more resources in human capital development and in creating an enabling environment for women and girls. There are compelling economic reasons for engaging women more prominently. Closing the male-female employment gap is good for economic growth. Research indicates that the narrowing in the male-female employment gap has been an important driver of Europe's economic growth in the last decade. In the Asia and Pacific Region, for example, restricting job opportunities for women is costing the region between USD 42 and USD 46 billion a year. World Bank findings demonstrate that similar restrictions have imposed massive costs throughout the Arab States Region, where the gender gap in economic opportunity remains the widest in the world today. The World Economic Forum reveals that those countries that are role models in dividing resources equitably between women and men, regardless of their level of resources, fare better than those that do not.

Engaging women and girls in ICT sector work is not only the right thing to do from the point of social justice. It is also smart economics. Gender diversity in high value ICT jobs in both management and on companies' boards is good for business performance. Studies exploring the link between women in leadership positions and business performance have shown a direct positive correlation between gender diversity on top leadership teams and a company's financial results. More diverse teams make better informed decisions, leading to less risk-taking and more successful outcomes for companies.

A combination of approaches that ensure that more girls and women benefit from pro-women policies and are prepared for the future workforce underscores the need for training and career support at three distinct levels: for entrance levels by way of education, training, recruitment, internship and career incentives – which require a national reassessment of educational infrastructure and delivery systems; for mid-career levels through career promotion and training; and for management and senior levels through mentorship, up-skilling and sponsorship programs;.

At the same time parents, teachers, career guidance counsellors and recruiters need to shift their own mindsets to acknowledge that ICT careers are an important and viable opportunity for girls. And in order to secure initial gains made, women already active in the ICT sector need to take time to engage with community initiatives to mentor girls and young women and participate in virtual and face to face communities of practice.

The changing character of ICT occupations has intensified the need to ensure that graduates emerge with skills that match employer demand. These demands shift away from traditional ICT occupations (such as computer programmers) and towards business/ICT specialists, highly specialized ICT areas and multidisciplinary ICT occupations. This puts increased pressure on educators and the sector to guide interested students into relevant ICT education and career paths.

The ICT sector has changed radically since early computing days – and the 'knowledge economy' is now taking on hitherto unseen dimensions where communication technologies have become forces of social change. Social media and its participatory formats are as much about the technologies as they are about their applications – bringing the virtual and physical worlds closer together in dynamic ways across several platforms.

The development of new goods and services is expected to drive demand from businesses, households and governments; with replacement ICT investments further boosting continuing demand. Much of the growth of the highly globalized ICT sector comes from the efficiencies gained from the global re-organization of research, development and production to provide new and improved ICT products and services to new and expanding markets. This includes the expanding use of software and extensive application of outsourcing. Additional ICT growth is expected to come from "green growth" through "smart" applications in buildings, transport, energy, and production which will translate into demand for customized applications.

As ICTs merge with sector-specific technologies across the economy, they produce "hybrid jobs". The expectation is that young women will show more interest in opportunities that use their creativity and intuition, in for example software application design. Their future is particularly promising in bioengineering, power grid informatics, digital media, and social and mobile apps; these are interesting, fun, creative and social mashed-up hybrid jobs that combine ICT with business of every imaginable field.

## References

Carnoy, Martin. (2005), *ICT in Education: Possibilities and Challenges*. Universitat Oberta de Cataluny.

Grossman, G. and E. Helpman (2005), "Outsourcing in a global economy", *Review of Economic Studies* 72: 135-159.

International Telecommunication Union. (2009). *Measuring the Information Society: The ICT Development Index*. pp. 108.

Oliver, Ron. (2002). *The Role of ICT in Higher Education for the 21st Century: ICT as a Change Agent for Education*. University, Perth, Western Australia.

Walter Ong, (1988). *Orality and Literacy: The Technologizing of the Word.* London, UK: Routledge.

# 3

# Computer-Mediated Communication (CMC)

Computer-Mediated Communication (CMC) is any form of communication between two or more individual people who interact and/or influence each other via separate computers through the Internet or a network connection. CMC does not include the methods by which two computers communicate, but rather how people communicate using computers. The way humans communicate in professional, social, and educational settings is different, depending upon not only the environment but also the method of communication in which the communication occurs, which in this case, is through the use of computers.

CMC mostly occurs through e-mail, video, audio or text conferencing, bulletin boards, list-servers, instant messaging, and multi-player video games. More recently, weblogs and wikis have come to provide interesting alternatives.Switching communication to a more computer mediated form has an effect on many different factors: impression formation, deception and lying behavior, group dynamics, disinhibition, and especially relationship formation.

CMC is examined and compared to other communication media through three common aspects of any forms of communication: synchronicity, recordability, and anonymity. Each of these aspects vary widely for different forms of communication. For example, instant messaging is highly synchronous, but rarely persistent since one loses all the content when one closes the dialog box unless one has a message log set up or has manually copy-pasted the conversation. E-mail and message boards are similar; both are low in synchronicity since response time varies, but high in persistence since messages sent and received are saved. Anonymity and in part privacy and security, depends more on the context and particular program/web page being used. It is important to remember the psychological and social implications of these factors, instead of just focusing on the technical limitations.In a study conducted by Dr. Jeffrey T. Hancock, director of the Computer-Mediated Communication Research Laboratory at Cornell University, deception in commonly used communication mediums, the telephone, email, and instant messaging, was examined to determine how communications technology affects lying behavior relative to Face-to-Face (FtF) interactions.

The three design aspects studied were ''synchronicity'' (the degree to which messages are exchanged instantaneously and in real-time) of the interaction, the ''recordability'' (the degree to which the interaction is automatically documented) of the medium, and whether or not the speaker and listener are ''distributed'' (they do not share the same physical space). What was found in the study was that synchronous media increase opportunities for deception.

In a FtF, instant messaging, or telephone conversations, people lie more often because most lies are unplanned. In a less synchronous medium, such as email, the opportunity for spontaneous lying is much less, thus lies occur less often in email. Another finding of the study was that users will be hesitant to lie in a medium which can be recorded and reviewed. Instant messages can be logged by services such as DeadAIM and [[AIM+]], third-party add-ons to the instant messaging program AOL Instant Messenger, emails are saved by both sender and receiver.

Users are more likely to lie when using FtF or telephone interactions which are typically not recorded in order to avoid being caught. Lastly, participants restrain from lying if the media is not distributed because being copresent limits deception concerning issues contradicted by the physical setting. The example from Dr. Hancock's report is "I'm working on a case report" when in fact the speaker is surfing news on the web. The more synchronous and distributed, but less recordable, a medium is, the more frequently lying should occur. Of the four mediums studied, lying occurred most frequently on the telephone, followed by FtF and instant messaging, and least frequently in e-mails.

## Modes of Computer-mediated Communication

Computer-mediated communication refers to human communication via computer. The emphasis is on interåction between humans using computers to connect with one another. The computers may be central repositories for human messages, or they may comprise a network of links and nodes facilitating the transfer of human messages. Various modes and media can be combined to facilitate the communication process.

Presently, most applications and much of the research into CMC is focussed on text-based communication. With the increasing sophistication of computer and communications technology, however, additional modes and media are being utilized. Ultimately, in support of human communication via computer, "the human intellect and sensorium are entitled to the fullest array of media we can muster". Further, computer technologies that support human intellectual activity as well as mediate communications should be given appropriate attention in the realm of CMC as their power and usefulness increases.

There are numerous incarnations of text-based CMC. Foremost among the asynchronous types is computer conferencing (CC). CC usually refers to those systems that reside on large central computers to which numerous users can connect directly via inhouse terminals or indirectly via home computers, modems, and telephone lines. Messages are composed and saved to files that are designed to permit multi-user access.

As such, groups of people can establish "conferences", or "special interest group" (SIG) areas to which they send the messages they wish to share with others. In this manner, the "conference" resembles an ongoing discussion or meeting that does not require the physical presence of the participants in one place. Nor does it require that contributions to the discussion be synchronized with the attendance of other members of the group. All members can link into the conference discussion when they log in, at their own convenience.

Electronic mail (EMAIL) is another form of text-based CMC that has seen tremendous growth in business organizations. With EMAIL, there is no group-accessible central repository of messages. Rather, messages are sent to individual addesses just as conventional mail is sent to individual addresses. Of course, bulk mailings by individuals or by automated message handling software are becoming as common as their counterparts in the conventional mail systems. One variation on this theme, which is popular in higher education networks, is the LISTSERVER. Listservers function primarily to receive messages from individuals and then to distribute them according to mailing lists. Members of these mailing lists, then, constitute a group of communicants who can carry on asynchronous discussions much the same way as users of computer conferencing systems do.

Electronic bulletin board systems (EBBSs, or BBSs) are an emergent form of CMC activity. The original concept was that of the conventional community bulletin board as an area where public messages could be posted and read. In the early 1980s, BBSs were primarily established by computer user groups for the purposes of online distribution of software and related information. The messaging feature became increasingly popular as members found that the posting of bulletins soon resulted in interactive discussions.

That is, bulletin messages could either be initial postings, or response postings. The group discussion potential of BBSs led to a broadening interest in the medium. With the advent of post-hobbyist home computers and the availability of inexpensive communications software, BBSs by and for the general public emerged. Many of these new BBSs were no longer oriented to computer enthusiasts. Instead, they came to be used as local fora for educational activities, special interest groups, writer collectives, and academic discussions.

One popular feature of BBSs and of the larger commercial conferencing systems is the "CHAT" mode of interaction. This feature is conceptually based on another more conventional public communication system: Citizen's Band Radio. On a small BBS, users have the option of requesting a real-time, text-based chat with the SysOp. On larger systems, numerous simultaneous users can interact in real time and all participants can observe the contributions of the others as they occur. Just as the radio waves provide for multiple simultaneous users, the CHAT mode of large conferencing systems allows for similar interaction in text.

CHAT mode is an example of synchronous text-based CMC. Most computers used for communications have at least a rudimentary form of this feature installed. The various incarnations of synchronous CMC differ primarily in the way the exchange is presented on the screen, and in the number of simultaneous users who can participate in any one group activity. The exchange can be character by character, line by line, or in multiple-line blocks. In character by character mode, the recipients see each character as it is typed by the sender. In line by line mode, the message is not sent until the sender presses the key (or ), and in the multi-line blocks mode, several lines are composed before the message is sent. In all cases, however, the transaction occurs immediately and is not held in a mail system or conferencing space for future retrieval. This is not to say that transcripts of the interaction cannot be collected (logged) for future review of the discussion; it is just that the exchange is intended to be in real time.

The user interface for synchronous text-based interaction often involves some partitioning of the terminal screen into window areas. A common design for two simultaneous users involves splitting the screen horizontally. If the top half is used for receiving messages, the bottom half may be used for composing and sending messages. Systems with more advanced GUI (graphical user interface) features can allow for multiple windows of variable size and location to appear on the screen with any or all of them being used to receive or send messages. The CSILE Phone utility developed for the synchronous dyads in the present study used one window for both receiving and sending messages character by character. Other window space was reserved for future uses such as review of previous transcripts or access to information files.

### Asynchronous Computer-mediated communication

*In The Third Wave,* Alvin Toffler documented many of the current trends in society and projected future scenarios. His discussion of "flextime" included the following:

> Indeed it is the computer which has made flextime possible ..., but it also alters our communications patterns in time, permitting us to access data and exchange it both 'synchronously' (ie: simultaneously) and 'asynchro-nously'.
>
> What that means is illustrated by the growing number of computer users who are today engaged in 'computer conferencing'. This permits a group to communicate with one another through terminals in their homes or offices. Some 660 scientists, futurists, planners, and educators today in several countries conduct lengthy discussions of energy, economics, decentralization, or space satellites with one another through what is known as the Electronic Information Exchange System [EIES]. Teleprinters and video screens in their homes and offices provide a choice of either instant or delayed communication. Many time zones apart, each user can choose to send or retrieve data whenever it is most convenient. A person can work at 3:00 A.M. if he or she feels like it. Alternatively, several can go on line at the same time if they so choose.

EIES was designed by Murray Turoff and had its test period in 1977-1978. Turoff also designed and implemented the very first computer conferencing system in 1971. It served as an automated Delphi conference for the Office of Emergency Preparedness in the Executive Office of the President of the USA. Starr Roxanne Hiltz, a sociologist, began reporting on the effects and impacts of computer conferencing in 1975. Together they wrote the now classic The Network Nation which remains important because of the breadth of analysis and insight it delivers. Hiltz and Turoff defined, clarified, or predicted many of the issues and problem areas in CMC that remain as the focus of questions and investigation more than ten years later.

The wave of the future in human communication, however, is more than Toffler's characterization of "flextime" as a simple matter of convenience. In fact, asynchronous CMC (computer conferencing) was propelled to the leading edge of a new communications revolution primarily because of the way it alters human communication patterns.

As early as 1975, with relatively unsophisticated conferencing software (FORUM), Jacques Vallee, Robert Johansen, and Kathleen Spangler of The Institute of the Future were able to report on computer conferencing as an altered state of communication with significant social implications. Their report presented many of the same factors and impacts that continue to be discussed in the more recent CC literature. Three main themes or issues predominate: alterations in time and place, text-based interaction, and socioemotional content.

Turoff is particularly concerned with the use of computer conferencing for group support. He contends that the asynchronous mode of interaction is most important. This is not necessarily because of time and place convenience, but because "the potential for real improvement in the group process lies in the fact that individuals can deal with that part of the problem they can contribute to at a given time, regardless of where other individuals are in the process". The result, he claims, "is to free the individual to deal with the problem in ways consistent with his or her cognitive style".

It appears then, that Turoff is discussing a particular type of problem or decision task, or perhaps a certain phase in the problem-solving or decision making process. This phase would be one that involved exploration of possibilities and ideas for general consideration. It would have to be a phase where the consensus of the group would not yet be necessary, since once consensus is needed it becomes more important for individuals to synchronize their contributions with those of other individuals in group activities. Turoff suggests the software should be designed to synchronize the asynchronous interaction of the group.

With regard to text-based interaction, Rapaport states that it "addresses a potentially wider set of domains than any other teleconferencing or office automation system. Its purpose is the association of ideas and experience of people, communicated in written words". Feenberg, in his paper "The Written World" adds, "A group which exists through an exchange of texts has the peculiar ability to recall and inspect its entire past". In these two statements we have the makings of a natural hypertext: the association and recall of ideas in text. With the aid of computer based utilities to facilitate the linking and association of ideas, we can presumably make a quantum leap beyond any bandwidth constraints on text-based communication. Besides, whereas a picture is said to be worth a thousand words, one might also proclaim that certain combinations of words can elicit a wealth of images, sounds, feelings, and other sensations.

Provenzo refers to the proliferation of electronic text and information networking opportunities as the "electronic scriptorium" and he predicts that:

> As we go beyond the Gutenberg Galaxy and enter a post-typographic culture, we will see the traditional methods and results of scholarship redefined. Through the establishment of information networks and knowledge pathways made possible by the microcomputer and telematic systems, we will have the potential to extend and refine intellectual discourse in a manner that has never been possible before.

Considering CMC in the context of the evolution of media, Levinson describes the development of communications technologies in three historical stages. First, there is the stage of natural communication that takes place within the confines of biological boundaries. The next stage involves the development of technologies which help transcend the physical limitations of human biology. Writing is a technology of this second stage. It evolved in order to facilitate communication across greater distance and through time.

According to Levinson, however, new communication technologies in the second stage often result in the sacrifice of certain features and qualities of communications in our natural environment of the first stage. The written word, and ultimately the advent of the printing press, resulted in the loss of interactivity in our communications. In the third stage, technologies evolve in a manner that recaptures features of natural human function; a process that Levinson refers to as "anthropotropic". Therefore, computer conferencing is proclaimed as "the first progress in the evolution of media since the mass application of the printing press in fifteenth-century Western Europe" because it recaptures the lost interactivity in text.

Without encompassing the full range of human sensory and expressive capabilities, text-based interaction is often thought to be an impersonal medium devoid of social context cues and nonverbal communication. Experience and research, however, are demonstrating that socioemotional content can be communicated in text. Steinfield states that: "Evidence continues to mount showing that CMC will be used for emotional interaction. People seem to work around the nonverbal cue limitations and actively provide their own text-based translations of nonverbal cues".

In terms of cognitive effects, very few studies address the issue directly. Most report on social relations, interaction and outcomes of problem-solving tasks. Kerr and Hiltz, however, surveyed the most prominent CMC projects of that time by soliciting opinions and observations from key representatives of the various groups and organizations utilizing this technology. In line with much of the research preceding their study, and much that would come after, a fairly standard set of factors and effects were analysed. These included system software design factors, and factors affecting user acceptance of new systems.

Kerr and Hiltz also gathered information about the cognitive impacts of CMC on individuals and groups. They reviewed the existing literature, and used reports from evaluators to confirm or disconfirm their expectations . They found that, "the more socially significant cognitive impacts, such as those including conceptual skills and learning [were confirmed], whereas those which may be more trivial, such as spelling and typing skills, and those which are clearly negative in impact, such as information overload [were not confirmed in the literature and reports]".

## Synchronous Computer-mediated Communication

Most computer supported communication systems have some capability for synchronous text exchange. Strom notes, interestingly enough, that the PARTY-LINE system developed by Murray Turoff was the first to implement a synchronous exchange facility. Rapaport confirms this and goes on to state: “Initially, synchronous support was expected to provide the system’s primary value. Dr. Turoff quickly discovered, however, that most of the real work was done asynchronously”. Hiltz and Turoff also state: “Of all the communication forms and conditions permitted by computerized conferencing, the synchronous discussion seems to cause the most difficulty and feelings of confusion” . The evidence provided by Hiltz and Turoff is rather slim. They relate the experiences of people in one real-time “party” where insufficient computer core memory allocation resulted in line by line messages taking up to 30 seconds to transmit. There was no structuring of the communication protocols. In fact, there was overt refusal by Roxanne Hiltz, who initiated the “party”, to take a leadership or moderating role.

Dobos and Grieve studied the decisional productivity of synchronous online conferencing by comparing the ratio of decisional output messages to messages representing social presence, task related input, and group procedural input. Their results suggest that the use of turn-taking protocols and the use of rotating moderators (with each participant “handing-off” to the next when their contribution was finished) increases decisional productivity. Further investigation demonstrated increased participant satisfaction when turn-taking protocols where implemented in online synchronous groups.

Kiesler and co-workers at Carnegie-Mellon University investigated the social psychological aspects of computer-mediated communication. Their 1984 paper was published in American Psychologist, became very popular, and is often cited. It was also selected for inclusion in an important collection of papers on computer-supported cooperative work four years after its original publication. In fact, it is a very well presented report of research that carefully delineates the factors and effects studied.

The problem with this account, however, is that they failed to clarify one very important point. Whereas they used the terminology of computer-mediated communication, they did not emphasize the fact that most of their data was drawn from synchronous text-based interaction, not asynchronous computer conferencing. Further, although they did present a portion of their work comparing users of the synchronous mode with users of electronic mail, they continued to confuse the issue by referring to their synchronous mode as “computer conferencing.”

The effect of this combination of a popular scholarly report and the confusion in terminology relevant to the field, is that some of its main findings have been misapplied to the more widely implemented and researched asynchronous mode of computer conferencing. Most significantly, the issue of uninhibited verbal behavior, reflected in the fiery term "flaming", came to be associated with CMC in general.

The truth of the matter is that the data presented in Kiesler et al. show higher levels of uninhibited verbal behavior in synchronous CMC, but the levels for EMAIL (an asynchronous mode) are more comparable to the lower levels found in face-to-face interaction.

An important new perspective is offered by Ferrara et al. who refer to synchronous CMC as interactive written discourse (IWD) and suggest that it represents an emergent linguistic register. The concept of "register" deals with language variations in the context of language use, as opposed to "dialect" which refers to language variations according to user. Ferrara et al. studied synchronous text-based CMC in the form of message exchanges between a travel planning agent and 23 computer professionals (or their spouses) who were making travel arrangements. Their findings support three claims about IWD:

> ... first, that it is a naturally occurring register, perhaps a reduced register; second, that it is a hybrid language variety, displaying characteristics of both oral and written language; and third, that norms of its use are in the process of becoming conventionalized.

"Reduced register" refers to registers that frequently omit components of the language such as copulas, articles, and pronouns. Note-taking is a reduced register that is generated under real time conditions with pressure on the writer to save time and minimize effort. Ferarra et al. point out that IWD is a register comparable to Note-taking in the type of register reduction it exhibits; except for one crucial difference. IWD is interactive and there is need for somewhat improved clarity in the communications because the audience is other than the self.

## Computer-Supported Cooperative Work (CSCW)

Most human work involves both individual effort and group participation. The "personal" computer and common applications software such as database managers, wordprocessors, and spreadsheets have provided significant support to people working independently in various organizations. These people, however, must step out of their computer support environment whenever group activities such as meetings are scheduled. In response to this problem, and in conjunction with developments in media and communications technologies, a new area of research and practice has evolved. "Over the last half-dozen years, Computer-Supported Cooperative Work has emerged as an identifiable research field focussed on the role of the computer in group work".

CSCW has drawn together research interests from a variety of fields in order to investigate groups of humans using groups of computers in their work. Sociology, anthropology, and organizational behaviour offer methods and knowledge about the study of humans in groups. Computer science offers methods and knowledge concerning computer-human interaction (CHI), interface design, and communications networking. Together, the integration of these perspectives have resulted in the development of applications that provide research information about group work in shared multimedia spaces. The general direction presently suggests the use of advanced multimedia technology to facilitate group interaction concerning the shared objects of work.

The strength and importance of this trend is revealed in the fact that the latest operating system and user interface for the Macintosh computer incorporates shared document and workspace capabilities as a standard feature of its networking environment. Tocco Gives an example of the use of System 7 features for group work. With the combined effort of several people on a shared document, he states that the document seems to come alive as it evolves with the completion of each participant's contribution.

It appears that the time for individual, "personal computing" software applications is fading away (at least in organizational settings). Document sharing and computer-mediated discussions in support of group work are on the rise. The term "groupware" is being applied to various commercial applications packages that integrate facilities for document sharing and messaging with the more common text-processing, database, and spreadsheet functions. Greif declares,

> All software will be groupware. But CSCW as a research field will still be addressing the larger questions of how to design and refine good groupware - software that will allow people to work together with the best help they can get from the computer.

A related development in the field of management science is the work involving group decision support systems (GDSSs). This represents a variation on the CSCW theme. After the groupware model, GDSSs attempt to integrate group sharing and communication structures with management software. In particular, GDSSs are best understood as the groupware version of more conventional decision support systems (DSSs). DSSs use operational data from an organization to provide simulations and models to help inform management decisions.

Baecker provides a taxonomy of CSCW technology that classifies applications according to synchronicity on one dimension and geographical dispersion of the group on the other dimension. He classifies computer conferencing in the asynchronous, multiple-site cell; electronic meeting and decision rooms in the synchronous, single-site cell; and media spaces, telepresence, and synchronous groupware in the synchronous, multiple-site cell. It is interesting that the cell for asynchronous, single-site applications is left blank, because it is certainly possible to have a messaging system or a group note-making system located at one site but used intermittently by the members of the group as an alternate mode of communication for certain tasks.

Computer conferencing, for example, is often used at university sites as a supplement to face-to-face discussions held in the classroom. Similarly, online courses that utilize CC have participants working from the same site as well as distributed locations. In another example, CSILE establishes a variation of asynchronous, single-site group interaction as the students of one classroom can enter notes, as well as review and comment on the notes of others in their class.

The concept of a single-site asynchronous utility may represent an area of knowledge and interest dissonance between proponents of CSCW and proponents of CC. In fact, the blank cell in Baecker's taxonomy may indicate a lack of appreciation for the asynchronous text-based interaction promoted by Turoff and others. Turoff on the other hand, is quick to downplay the importance of GDSS and CSCW by suggesting that computer conferencing already offers GDSS via a flexible, tailorable system and that CSCW, being mostly synchronous in nature, cannot match the group support power of asynchronous CC. It should be clarified as well, that Turoff has good grounds for his view that work in CC has already addressed a lot of the issues now arising in the GDSS and CSCW work.

## Infrastructural Issues

A fully networked society, with capacities for one-to-one, one-to-many, and many-to-many communications, provides a fertile incubator for the development of next-generation information services and for the development of next-generation information workers. Many facets of the infrastructure are already in place, however, the parts remain scattered across networks, operating systems, and computing platforms.

Computer communications can be used to simultaneously join multiple users across hundreds of networks. Recent advances in networking technologies enable diverse computer, information, and telecommunication systems to co-exist and integrate on the networks. While there has been a dramatic increase in the use of wide-area networking technologies, these systems have yet to achieve a full integration into organizations and society.

If fully implemented, network applications might dramatically impact intra- and inter- organizational communications to improve company operating procedures, create strategic advantages, or forge new business alliances. Internet resources have recently come within reach of most organizations and computer users. Recent advances in the technologies of regional, long distance, and value-added telecommunications carriers have similarly created new opportunities to enhance communications. Organizations and individuals may connect to the infrastructure through public, private, or value-added network providers.

Business, industry, government, and education can all claim adoption of sophisticated information technologies able to adequately serve internal operations. Effective organizational communications increasingly involves the integration of resources distributed across organizations, and networks, with the objective that users be able to share all information within a total information environment. External communications for trade, research, and intellectual pursuits are becoming increasingly important. However, effective applications of wide-area, inter-organizational communications are still relatively rare.

### Electronic Environments

The goal of telecommunication planners is to create a global digital highway, open to all telecommunications users, with universal access into totally open networks. Users will be able to economically access voice, data and images, in any combination, anywhere, and at any time. Such capabilities exist today.

However, effective use of the resources has too often become dependent on information policies and the culture of the organization. Adoption has entailed a rather difficult process of unscrambling old procedures and attitudes, moving to new ways of performing intellectual tasks and of thinking about communications, and then installing the new processes into the daily agenda of individuals and groups.

While 83% of U.S. Government workers would prefer to work via computer at home or at a near-home remote office, only 14% have the opportunity. Organizations have yet to decide whether office technology has produced measurable increases in productivity, or even whether traditional productivity measures adequately assess quality and quantity of work. Strategists implementing information projects must consider not only systems development needs, and the required skills of employees, but also the degree to which information services are supported by the organization.

Traditionally, implementation problems have been caused by differing interface standards and modes of human-computer discourse. Fortunately, graphical user-interfaces have somewhat alleviated these concerns by standardizing screen displays across multiple platforms, networks, and operating systems. In addition, new network applications allow intelligence to be programmed into the user-interfaces to enable systems to maintain and apply information about the user to further ease communications.

**Online Communities**

Online communities that will be instrumental in the realization of an advanced learning society. Within these communities, users will access remote computer systems and resources which are fluid, evolving, and continuously expanding. The resources may be used to improve communications, stay abreast of current developments, maintain contact with business partners or external specialists, and to establish future business opportunities.

Difficulties arise because inter-organizational communications services are often not immediately recognized as critical for business functions. Future 'opportunities' are sometimes difficult to justify, thereby hindering wide-spread adoption. Thus, while technology may progress rapidly, the human processes to successfully implement and employ that technology may not.

Implementation problems are compounded because of the somewhat unusual nature of 'networking.' Namely, with network technologies, information flows throughout organizations, and among different organizations, freely and without regard to corporate interests. Some argues that the ultimate impact of the new technologies is to give end-users greater power to shape their computing systems and to manage their information needs. This is very different from traditional 'top-down' communication patterns and can cause concern for top management. Yet, it is this empowerment of the end-user, and their interaction with fluid and dynamic resources, that will drive future organizational innovation.

As users acclimate to network-based communications and information access, they also become increasingly aware of data organization and retrieval issues, and are thereby introduced to current practices in database design, development, and maintenance. Thus, computer-mediated communications may evolve the culture of the organization as the workforce becomes more adept at capitalizing on institutional information investments to increase economic productivity.

**Human-Network Issues**

The primary modes of computer-mediated communication have been classified as either immediate (teleconferencing), delayed (correspondence), or simulated (computer-tutorial). In addition, new 'groupware' applications are a natural outgrowth of computer communications and enable multiple users to simultaneously interact on the same computer project at the same time.Groupware applications can improve inter-organizational communications, be used for in-service training, and can improve overall productivity. Employees will need training in the use of the networks. Fortunately, online systems are increasingly adept at supporting in-service training and education. National mailing list services can support network training and alleviate a need for internal funding.

Online information may also be customized to satisfy the needs of participants. And, throughout the development, implementation, and training process, there is little need for administrative costs as users at desktop workstations manage their personal communications, resource sharing, and business transactions.

Network communications are pervasive, reaching into the daily activities of users wherever they work, live, or attend class. Users may access the network at a variety of times, from a variety of places, and participate for a variety of reasons. As this convenience is realized, electronic technologies will increasingly influence program-planning and development strategies as representatives of the target population become more deeply involved in information development and presentation.

In addition, computer communications are dynamic, enabling current knowledge to be integrated into organizational communications. As employees share new knowledge and information, they garner new perspectives and thereby drive innovation. Throughout this process, the medium can itself be used to coordinate the creative contributions of the individuals.

**Infrastructure Development**

The Internet is excellent for wide-area data transfer and automated online services, but poor for real-time multimedia communications. Private common-carrier networks provide viable multimedia solutions, but offer little, if any, public information. Value-added networks offer some solutions for both Internet and multimedia communications, and some provide online resources. An effective infrastructure may require a mix of networking technologies and services.

In areas of the country with an abundance of high technology companies, such as California's Silicon Valley, many employees have Internet access at the desktop. However, they are not active users. The University of San Francisco, College of Professional Studies, is actively involved in educational programs for mid-career information systems professionals in Silicon Valley. The following network overviews have proven helpful when discussing the roles of infrastructure service providers.

Networking has become a national priority. Compatibility and interoperability standards are established under the Government Open Systems Interconnection Profile. The National High Performance Computing Technology Act will evolve the Internet into the National Research and Education Network (NREN). NREN will extend network technology into homes and organizations, establish comprehensive online information services, and create opportunities for partnerships among government, education, and business. Organizations may connect to the National Science Foundation backbone through one of the regional public networks.

Participation often depends on the needs of the organizations, the aggressiveness of the local sales force, and company motivations.

The Silicon Valley area has several public and private Internet providers, and many of the high technology companies themselves make networking products. The BARRNet regional Internet network, in addition to connecting the universities, has active participation by over 70 prominent Silicon Valley high technology companies. However, while Internet communications with employees within these companies is possible, and even somewhat common in certain situations, the Internet is not being used to its fullest potential. Individually, interest among employees is very high, especially once they have seen some of the capabilities of the network.

Internet information services are a valuable resource and solve many of the training and acclimation problems inherent in learning the Internet. Fortunately, such training is available, free-of-charge, through mailing list services, and as self-contained instructional modules for downloading from public domain facilities. Useful training media, such as BBN's 'Internet Tour,' and Merit's 'Navigating the Internet,' have proven quite popular with information systems professionals. Downloading these resources from anonymous ftp sites, using a graphical retrieval and translation utility is also a useful experience.

'Telnet' exercises in which users login to remote computer systems and evaluate national and international library, public information, and government computing resources serve to teach students about network database structures. The Hytelnet library navigation program is a useful telnet demonstration utility. Subscriptions to mailing list services, such as NETTRAIN, Interpersonal Computing and Technology (IPCT), and Comserve, can help organizational innovators learn more about the Internet.

For more advanced users, demonstrations of the new network access, retrieval, and processing utilities impress even the most seasoned information systems professional. A Wide-Area Information Server (WAIS) search can illustrate the power of wide-area network database systems by retrieving information from distributed databases using a natural language query. Internet 'Gopher' software can be used to demonstrate a technique to search and display files structures from remote sites. A 'World-Wide Web' hypertext search can aid research in network-based, human-machine interaction.

In the private sector, new telecommunications switching technologies are melding previously incompatible systems and connecting diverse networks and computing platforms. This is timely since multimedia communications are spurring a demand for faster, more efficient, and more compatible networking. A trend is clearly underway as modem users require more capacity and evolve into online X.25 packet networks. X.25 users are demanding multimedia and evolving into higher capacity services such as Frame Relay, Switched Multimegabit Data Service, and Asynchronous Transfer Mode (ATM).

Multimedia communications are increasingly driven by desktop conference systems able to integrate voice, image, and video into computer-mediated messages. Such multimedia communications can be built into the physical apparatus and wiring in systems such as the Integrated Services Digital Network (ISDN). Or, they can be software-based and achieve multimedia interactivity through desktop systems developed by Northern Telecom, AT&T, IBM, and professional multimedia publishers. Multimedia has become a major force advancing the technology of organizational communications. Another interesting opportunity results from the systems upgrades of the regional and long distance carriers to digital switching and SS7. These technologies enable users to program network functions on both a regional and national level. And, as the networks become increasingly 'intelligent,' programmable network applications will enable users to define personalized, intuitive, multimedia services customized to individual needs.

### Value-Added Solutions

Organizations may also connect to the Internet through several value-added network providers. Some of these services actually improve on some of the basic Internet communication capabilities. In conventional Internet switching, the TCP/IP data grams are encapsulated in X.25 or frame relay networks. The value-added providers switch TCP/IP in native format, using routers as switches, thereby eliminating switching delays. These providers include Advanced Networks and Services, UUnet, Performance Systems International, CERFnet, and Sprint. A liberalization and commercialization of the Internet, and public demand, have expanded opportunities for Internet access for individual users. Anyone with a computer and modem can now access full Internet services. Following is a list of the basic Internet functions provided by private online services and available for individual use:

| Service | Cost | Mail | Telnet | FTP | Netnews |
|---|---|---|---|---|---|
| AOL | $8 | yes | no | no | yes |
| CompuServe | $8 | yes | no | no | yes |
| Infoserv | $14 | yes | no | yes | yes |
| MCI | $5 | yes | no | no | no |
| Netcom | $18 | yes | yes | yes | yes |
| Prodigy | $15 | yes | no | no | no |
| Well | $15 | yes | yes | yes | yes |
| Delphi | $13 | yes | yes | yes | no |

All of the services offer mail boxes and Internet mail. Most are accessible through a local telephone call. In addition to the listed monthly subscription, hourly charges can vary depending on the service. Telnet and ftp are sometimes a minor additional charge. Of course, with the basic mail service, users can access the Internet listservs to find network training media. Usenet Netnews is also available through many of the suppliers. Most of these services provide bulletin boards and online databases so a subscription includes access to those services. Internet operations are a new but rapidly growing part of their services, and many of their databases contain Internet information.

The evolution to network-based information delivery seems a certainty. The success of the systems and programs will be dependent on the degree to which the processes are integrated into the information infrastructure of the organization and into the daily agenda of participants.

Wide-area network services may involve a mix of networking technologies and services. Too often, the academic community downplays the potential of private networks, and industry underestimates the potential of the Internet. Presently, organizations have needs for simultaneous information, conferencing, and multimedia communications which can only be satisfied through hybrid public-private wide-area network solutions.

Computer conferences, online databases, and bulletin board services afford users the opportunity to complete network training at their own pace and personal convenience. The media may be online text, or independent multimedia or hypermedia instructional modules. Group software, with members operating locally or at a distance, is an exciting area for exploitation and a natural outgrowth of online conference systems. Some of the more interesting aspects of the new network technologies derive from their potential to enable users to shape their personal information environments.

Future research may help to determine the proper mix of the various network and communication technologies, the impact of network-based communications on traditional information structures, and design processes for the distribution of information to various home, work, and social environments.

## Use of CMC in Education

Throughout the history of human communication, advances in technology have powered paradigmatic shifts in education. Technology changes both what we can do and what we decide is best to do; big shifts in culture cannot occur until the tools are available. The printing press is an example. Before its invention there were people who could read and write; yet not much reading and writing took place because, for one thing, books were costly and scarce. The press enabled widespread literacy, with books accessible and more affordable for all. The spread of literacy in turn changed both the educational system and the class structure, with consequences that still shape our attitudes today. When people began to accumulate knowledge through the technology of writing and reading, they found a way to preserve it through succeeding generations without relying on memory — greatly changing the way education was conducted.

The impact of the printing press on students of the time has been analyzed and reanalyzed. No longer did students have to write or remember everything the teacher delivered; students could use books. But they did not completely give up the oral/aural connection; witness the popularity of lecture classes even now. As new technology enables shifts at the level of delivery, old technologies are augmented, not totally replaced. Even though many of us have computers at our disposal, we still use books, speech, and pen or pencil writing in education.

The general availability of electricity has fostered an almost universal use of such inventions as radio, television, and, increasingly, computers. For decades, educational technologists have likened the impact of television and other electronic ways of presenting information to the impact of the printing press on learning. Although to date television has not had nearly the impact on school learning that books have had, we have yet to determine whether its impact on education as a whole will be as great and long lasting. Schooling is only a part of education.

Much of education takes place outside of schooling, both as planned activities and unplanned learning. We may not understand the instructional goals of the Music Television (MTV) channel broadcasts, and those goals may differ from those of educators, but that does not mean MTV viewers do not learn anything. Ultimately we must consider what kind of world we as educators want to build. If we envision the merger of computers and telecommunications as a new tool for teaching and learning, now is the time to clearly articulate and promulgate our goals in order to shape future uses of instructional technology.

For communications to take place, at a bare minimum, there must be a sender, a receiver, and a message. If this message is intended as instruction, then besides student, teacher, and content, we must also consider the environment in which this educational communication occurs — an environment that benefits the educational system in some ways and constrains it in others. Part of this learning environment can include various technologies and media. If "the medium is the message," that is, if technology changes what we can do and how we think about it, then the various media enabled by instructional technology also change both what we can do in education and how we conceive of it.

For many years, educators have been exploring ways to combine theories of differing learning styles and student- constructed knowledge with the theory of practice-centered learning. Instead of being passive recipients of knowledge, we now consider students capable of constructing their own knowledge with guidance from the teacher. We can offer part of this tutorial guidance by setting up an environment that will provide students with the resources necessary for independent exploration. In using emerging computer-based technology as a resource, students are encouraged to explore their own interests and to become active educational workers, with opportunities to solve some authentic problems.

As an agent for socialization and collaboration, the networked computer has an even greater potential in education than does the stand-alone, knowledge- server type of computer. The active environment of social learning provided by a computer with access to local, national, and international networks increases interaction and communication among students, their teachers, peers, parents, and other members of the world community. Although there are some differences between distance education and classroom education, the significant issues concerning the use of computer networking and other emerging technologies to promote learning in both are similar.

**Distance Education**

In addition to being entertained, viewers seem to learn from the Music Television (MTV) channel's eclectic mix of music videos, news, instruction, and information on the world of popular music and performers. Broadcast at a distance, MTV's educational content appears largely unplanned in the sense that educators are not directly involved. However, the distance education now originating from universities and colleges attempts to plan educational content and activities for students removed in place and time from their instructors.

Historically, we have not done a very good job of implementing the concept of learner-centered education in distance education. As Thornburg points out, it is difficult, at best, to instill a mindset of lifelong learning in others if we do not understand it and demonstrate it ourselves. One of the reasons that we have failed in this area has been that the tools were not available to do much besides deliver education (as opposed to enable learning) at a distance. Now, computers and telecommunications have opened the way to formats other than pen- and-paper correspondence courses and allow for a more interactive, integrated learning environment.

The type of change enabled by computer-mediated communication (CMC) does not just involve adding new technology to old ways of organizing teaching and learning . Although the perennial problem is still one of instructional content and design, we must not pave over old cow paths.

Another notion current in educational circles is that we need to develop motivated, skillful, lifelong learners. As knowledge in many fields increases exponentially, we cannot hope to fill up students as if they were passive, empty vessels. During formal schooling, aspiring professionals can only begin to take in the amount of information that they will need during their career life times. The knowledge base of certain fields may have appeared static for decades, but we can no longer accept that view. Therefore, we must teach students to become lifelong learners by helping them locate the resources to continue learning.

Distance educators are now beginning to focus on a related set of notions: (a) there are different learning styles, (b) students create their own meaning when learning new things, and (c) what makes a difference in content retention and transfer is not so much what is done by teachers, but what students as learners can be encouraged to do themselves.

Much has been written about the importance of accommodating the learning styles of different kinds of students. Suffice it to say here that too often students have little choice in what to learn, how to learn it, or when to learn it. The body of literature on constructivism which has emerged over the past few decades has also contributed to our understanding of learning styles. When content is meaningless to the students' world view, when they are taught as if they were passive recipients of knowledge, or when they have little engagement in the instructional tasks, students have no incentive to construct their own knowledge and little motivation to retain information or transfer its use to novel situations.

The notion of practice-centered learning (PCL) is also important to distance learning. As we learn more about how learning occurs, it becomes increasingly clear that the educational process takes place in a complex internal and external environment. One of the teacher's roles is to become the creator of an effective external learning environment that stimulates the environment within. How do teachers and developers of instruction create environments that are conducive to and enhance student learning?

The technology that can help provide these new environments for education is emerging. This technology allows us to utilize such methods as cooperative learning, to recognize such concepts as interdisciplinary needs in education, and to provide an environment in which collaborative efforts are rewarded. These methods foster a view of knowledge in which expertise is distributed and created among the different participants. Now there is no shortage of technology, only a shortage of the educational vision necessary to use the technology to create new educational environments.

## Use of Computers and Telecommunications in Classrooms

What we have been discussing is a reengineering of education, not only in the sense of rethinking the organization of site-based schools, but also in the sense of finding ways to unite computers and telecommunications and bring down the schoolhouse walls; to deliver instructional content when and where it is needed-whether in the home, the workplace, or the school.

Computer-mediated communication (CMC) promotes a type of interaction that is often lacking in the traditional teacher- based classroom. It allows learners the freedom to explore alternative pathways-to find and develop their own style of learning. What if content could be delivered in the form of graphics, text, and/or full-motion video, whenever and wherever in the world it is requested? How do we, as teachers and educators, responsibly participate in and make use of the inevitable technological changes at hand?

Computers are not a threat to the teacher (although the role of the teacher must change when using them), but computers may threaten the chalkboard. Computer technologies allow professionals to share with students tools that we use daily. Further, as educators, we can provide guidance to help students develop meaningful ways to construct their own knowledge, much as we ourselves do.

Technology enables us to implement these new visions in distance learning. Berge (in press) points out that: "[T]echnology makes it possible that these investigations are not limited to students from one classroom, school, grade, or country necessarily-nor to exclude experts in the field of inquiry from the collaboration. Effective learning hinges on active engagement by the student and the construction of knowledge on their own leads to understanding. This learning is not a solitary process. Rather, it occurs in a larger world of people and technology."

CMC and networking in general can promote long-distance collaboration among students and content specialists in many different areas. The integrated use of technology offers many educational opportunities and possibilities when driven by sound visions of learning. The students' ability to create knowledge can be enhanced when their instructors use varied instructional delivery formats to provide a richer environment than is used in most distance education practiced today.

However, as Sheingold points out, these ambitious new goals for student learning, along with radical changes in the students' roles those goals bring about, must be met with radical, ambitious changes in the educational process. Indeed, information technology offers options for reorganizing and refining distance education. But our new visions of distance learning must drive our decisions about our use of technology, not vice versa.

Although major cultural shifts do not occur without the tools that make them possible, once those tools are at hand, the shifts are inevitable. Emerging technologies, such as interactive television and the "superhighway" for information exchange, may look different depending on who builds them (e.g., telephone companies, cable television companies, federal governments), but we may be assured that they will be built by someone. How we as educators will participate in this enterprise is a issue that deserves our closest attention. More than merely a shift within education, our participation in this movement will bring about major shifts in society and culture. As the number of students outside the ages of 18 to 25 increases, and the number of persons requiring off-campus classes rises, the very existence and future of a university or college may hinge on serving this newly defined and diverse population. In this book, we hope to show how CMC can help serve that population.

In combination with other media, computers can utilize an instructional design that teaches to the multiple intelligences that Gardner speaks of in Frames of Mind (linguistic, logico-mathematical, intrapersonal, spatial, musical, bodily kinesthetic, and interpersonal). The idea behind this instructional design is to use as many methods and formats for instruction (e.g., small group discussion, graphics, lecture, hands-on labs, writing/ reflection, sound, CMC, and conferencing) as possible, provided that instructional goals and design dictate their use.

## Use of CMC in Workplace

The compter technology reinforces existing hierarchies within organizations, or at least does nothing to break them down. For librarians, especially those in the corporate environment, the question of whether technology may or may not democratize the workplace has much relevance for two reasons: As implementers of information technology, we may consider ourselves as corporate and social do-gooders helping to break down status and power-related barriers by providing access to information to anyone in the company. As employees, and thus users of new technology, we may believe that our voices have a further reach and that our input is taken more seriously.

Collectively known as computer-mediated communication (CMC), it includes e-mail, teleconferencing, electronic bulletin boards, Internet/Intranet access, Internet relay chats, and group decision support systems (GDSS).Lee Sproull and Sara Kiesler are among the most vocal proponents of the thesis that CMC has democratized the workplace. In their book, Connections: New Ways of Working in the Networked Organization, they argue that computer-mediated communication reduces information gaps by providing a tool for peripheral employees (employees that are removed from the core management structure) to be connected to all people within the organization. They examine passive and active connections in the CMC environment. Passive connections is where employees choose to be an observer only of messages being posted to distribution lists or e-mail. Active connections is where employees partake in the electronic interchange and discussions posted on listservs or e-mail.

The benefit of computer-mediated communication, suggest Sproull and Kiesler, is that it gives a "voice to the voiceless," and, therefore, increases "emotional and informational connections" among employees by creating more active participation in exchanging messages and establishing collaborations in electronic discussion groups. Giving people a "voice," the authors suggest, is part of our democratic culture. Information in organizations now flows more freely in the CMC environment and employees are more willing to participate beyond their stated duties.

Libby Bishop and David Levine, in their article "Computer-mediated Communication as Employee Voice: A Case Study", also examine the "employee voice through computer-mediated communication," supporting some of the findings of Sproull and Kiesler. They conclude that CMC creates easier access to people and information and provides a new vehicle for employees to address concerns to management in the hopes of resolving problems or effecting change. They examine the effects of CMC at a high-tech firm that was an early user and advocate of computer-mediated communication.

After collecting data and conducting interviews over a two-year period, their findings suggested that technology had eroded the hierarchical structure of management and gave employees more of a voice in the decision-making process of the organization. But has the impact of computer-mediated communication really democratized the workplace? Has it weakened or softened the hierarchical structure of organizations? Are librarians now free to interact and exchange information at all levels within or beyond the organization? And do librarians have a more "active voice" in the virtual environment in being part of the decision making process that they did not have in the past? Some researchers contend that CMC does not lead to a democratized workplace.

Susan Herring, in her study, "Gender Participation in Computer-Mediated Linguistic Discourse", was one of the first to challenge the position that modern information technology opens the communication path to a more egalitarian communication structure. Focusing on gender participation in the CMC environment, she observed two electronic discussion lists at an academic institution. She examined two issues for her study: 1) to see if the communication process was democratic in the electronic environment, and 2) to determine if computer-mediated communication increased gender equality.

Herring focused primarily on the degree to which males and females participated on these discussion lists. Her findings revealed that there was a difference between the female to male participation ratio. Male participation showed minimal interest in female postings, which resulted in a decline of female participation. Female postings to the discussion lists tended to be brief or ignored if certain subjects were not of interest or topics discussed did not encourage female participation. Herring concluded that "...although the medium theoretically allows for everyone with access to a network to take part and to express their concerns and desires equally, a very large community of potential participants is effectively prevented by censorship, both overt and covert." In this particular academic environment, computer-mediated communication was found to be male dominated, power-based and hierarchical.

Like Herring, Giuseppe Mantovani is critical of the egalitarian democratic effect of CMC. Focusing on a corporate, rather than an academic environment, he explores the conditions under which CMC may or may not promote democratic patterns. He challenges the arguments made by Sproull and Kiesler in his article, "Is computer-mediated communication intrinsically apt to enhance democracy in organizations?".

Much of how CMC is used depends on the culture of the organization, rather than just the technology itself. He suggests CMC is effective in overcoming physical barriers (geographic location), but not social barriers (structure of the organization). Social inequalities that exist in organizations are not solved by integrating technology. Computer-mediated communication, Mantovani believes, is determined by an organization's history and the rules that are implemented by management. Taking this into consideration, he says, CMC "does not generally foster democracy in organizations."

A more recent study by Frank Symons examines how traditional hierarchical structures are now reproduced electronically. In his article, "Virtual department, power, and location in different organizational settings", he says that technology gives management even more power than they had before by tracking production and efficiency with greater accuracy, less margin of error, and increased monitoring of employees' activities at work.

Hierarchy in the electronic realm depends on the degree to which organizational members have access to the Internet or company Intranet, electronic files or databases. To test the relationship between hierarchy, power, and location, Symons conducted a 10-year study that examined virtual departments in various organizational settings of firms with 3,000 to 5,000 full-time employees. Out of this research emerged three to seven different hierarchical categories, which were characterized by the complexity of the information that was accessible or shared among different departments. The question he raised was whether traditional hierarchies are really being eradicated by virtual departments, or are they merely being changed under the guise of technology.

Symons concludes that hierarchical models still exist in many organizations and are now just electronically reinforced. He notes that organizations practicing different management styles, such as centralized vs. decentralized, still show evidence of hierarchical structures because management holds the power in how the technology will be used, despite geographic location.

William Wresch, like Symons, supports the notion that management has power over technology. In his book, Disconnected Have and Have Nots in the Information Age, he discusses how information technology in many organizations has become the old "panoptican - an optical instrument that allowed an observer in the middle of a prison to see into all the cells yet remain unseen."

In the context of organizations today, this is how upper management controls the use and implementation of information technology - by the type of network administration software to be used, which is dependent on the server and connection an organization has, and who will administer and run these systems. Because of this control, Wresch states there is no "semblance of privacy" for employees anymore. Managers can physically be located anywhere and still have the ability to monitor and track organizational activities. He describes this control over technology as being very powerful for management since they now have tools to put employees under blind surveillance. The issue of power, technology and rights to privacy in the workplace is hotly debated today.

Studies supporting the democratizing impact of CMC are flawed. From the very outset, they fail to consider how the power structure in organizations controls how CMC is implemented and to what extent it is used or what restrictions are placed on it. There is an assumption that all information flows from top-down or vice versa, and that all employees are on the same level playing field as management in the virtual world.

Although technology provides us with a new method to transmit messages as senders and receivers, it does not ensure that these messages are heard or even considered. The reality is that technology, for the most part, has not changed behavior. In most companies, traditional power structures are still in place; messages can be ignored, and electronic activity is monitored.

Issues that confront librarians in an environment that is not democratized through CMC could include: e-mails ignored by upper management, monitoring Internet sites visited and e-mails sent, limiting access to information for certain employees, and information that should be posted in a visible location is buried under layers of directories and files that makes it difficult to find. Who has not personally experienced or heard from colleagues about suggestions repeatedly e-mailed to superiors that are ignored? Who has not encountered management's disapproval in one of its various forms from mild resistance to clear prohibition in response to an attempt to make company information accessible to all employees? And who has not heard about frightening reports of management monitoring Internet sites visited and e-mail traffic?

Technology itself does not shape the culture of an organization, as Mantovani states, although it is certain to have an impact on day-to-day transactions. There is no doubt that technology has changed how business processes and decisions are acted upon in organizations. However, the organizational culture is embedded by the vision of its leaders, and therefore, technology is adapted to "fit" into the culture—not the other way around. Computer-mediated communication has given employees a virtual illusion that there are fewer boundaries and less control placed upon them.

Computer-mediated communication is said to "enhance" and "empower" employees, but they cannot see how management exercises its power by controlling, maintaining and directing the use of technology. Under what conditions, then, can computer-mediated communication be a means in democratizing the workplace? Librarians need to be active participants in attempting to change the way in which existing power structures use technology.

Much of how we leverage knowledge and disseminate information will depend on the culture in which we work. Our desire to help in breaking down barriers is only successful if it is supported by the organization that employs us. In a setting where the power structure controls much of how the technology will be used and our input in the decision making process is not considered, our jobs as librarians is more difficult in trying to serve and meet the needs of our clients and, unfortunately, there are no quick fix answers.

What can librarians do in their organizations to realize some of the potential of computer-mediated communication for democratization within a more traditional hierarchical structure? The answer again depends on the organizational culture. One way to create more of a "voice" in a controlled and closed-communication setting is to establish a group or coalition of clients that support and value the services we provide for them. As a collective group, we have better leverage and can form a more powerful informal network that may have a greater impact in getting management to listen to our concerns. Another consideration in realizing the potential of CMC for a democratized workplace is that it will take time. More face-to-face communication is necessary in proving our value and stating our concerns or goals to management. Also, providing information to management illustrating, for example, the difference between abusing the Internet and using the Internet for information seeking purposes and better decision making may eventually open the door to more equal access to information for everyone in the organization.

## Potential Problems with CMC

Despite the advantages of CMC, there are some difficulties that are important to recognize. Often these difficulties stem from the same basic attributes as the advantages.

One of the most consistently recognized characteristics of CMC is the reduced number and variety of visual and auditory cues that are available. Some researchers, as we have just seen, feel that fewer cues encourage greater student participation. But others feel that the lack of cues can be associated with immature, insensitive, and unproductive behaviors. There is some indication that CMC results in more evaluative comments, including comments that are too critical. Participants sometimes retaliate in an insulting manner, and then the conversation can degenerate into what in the on-line world is called a flame war, an exchange of angry or derogatory remarks.

As a teacher, you should be aware that CMC messages can be misinterpreted and exchanges can become negative. Students involved in CMC can be sensitized to these limitations, learning to become more careful in the construction of the messages they send. Schools frequently provide students with a code of on-line conduct, often called netiquette.

## References

Haythornthwaite, C. and Wellman, B. (2002). The Internet in everyday life: An introduction. In B. Wellman and C. Haythornthwaite (Eds.), *The Internet in Everyday Life* (pp. 3-41). Oxford: Blackwell.

Herring, S. C. (2004). Computer-mediated discourse analysis: An approach to researching online behavior. In: S. A. Barab, R. Kling, and J. H. Gray (Eds.), *Designing for Virtual Communities in the Service of Learning* (pp. 338-376). New York: Cambridge University Press.

Markman, K. M. (2006). Computer-mediated conversation: The organization of talk in chat-based virtual team meetings. Dissertation Abstracts International, 67 (12A), 4388. (UMI No. 3244348)

Thurlow, C., Lengel, L. & Tomic, A. (2004). Computer mediated communication: Social interaction and the internet. London: Sage.

Walther, J. B. (1996). Computer-mediated communication: Impersonal, interpersonal, and hyperpersonal interaction. *Communication Research, 23*, 3-43.

# 4

# Digital Media Environment

Changes in media brought about by digitization and the Internet are much more fundamental and transformative. Yet ironically many of these fundamental changes will not be immediately apparent to the average media consumer. People will still watch movies and television; will still read books, magazines, and newspapers; and will still want to listen to music. For many people it will not matter that a song was created, produced, and even distributed digitally. A useful analogy in comparing the state of digital and analog media today is to picture an iceberg floating in the ocean. Just as only a small percentage of the whole iceberg is visible above the water, in today's media world the primarily analog media products we see, such as books or videotapes, are only a small part of the overall media creation process. Most mass media today already utilize digitization in some way, even if the final product is still predominantly analog. But this will change over time as consumers get more of their media content digitally. Four key concepts of digital media are:

1. *Multimedia*: Combining video, audio, and text is not unique to digital media-everyday television is just such an example of multimedia-but digital media allow for easier creation of multimedia than in analog media and provide for greater opportunities in fully integrating media types to complement each of their strengths.
2. *Interactivity*: The ability to interact with media content and obtain unique, personalized, or localized information is a powerful force in changing how the public uses and perceives media. It greatly shifts the

balance of power from passive media consumers to active media consumers and creators.

3. *Automation*: By creating various programs and automated functions such as search tools, collaborative filtering (automatically determining likely interests based on previously viewed or purchased material and comparing that to what others who have purchased the same material also bought or viewed), and updating content, computers greatly reduce the amount of work humans must do and, increasingly, can supplant some human roles in the media production workflow process.
4. *Ethereal quality*: Digital media are not actually physical products, like books, photographs or CDs, although they usually are eventually represented in some kind of physical product.

But even as far-reaching as these elements are in changing the media landscape, they are still incomplete without one important element: a network that connects computers or media devices to each other so they can communicate. There are many kinds of networks, but we will concentrate primarily on the Internet and the World Wide Web. Because all data that pass through the various networks-telephone, cable, or satellite-are at one or more stages digital, we will use the term "digital media" to mean not only digital but networked, or online, media as well.

Out of all modern media, the Internet/World Wide Web reached 50 percent of U.S. households faster than any other media technology. 58 percent of U.S. households, or 166 million persons, went onto the Web. The Internet is more complex in its requirements for adoption than all other media. One can't simply go out and "buy an Internet" as one could for most other media, such as radio or TV. Instead, one must first have a computer and a means of connecting it to the Internet, plus a higher level of technical media literacy to use the computer than is required for a television or even VCR. What Is Online Communication?

The broader topic of online communications, of which the Internet is a vital part, must be clarified. Although many may consider the term online as synonymous with the Internet, online is in fact a term with a larger meaning. Online refers to the interconnected, networked media that permit the direct, electronic exchange of information, data, and other communications. Everything from local area networks to wide area networks, such as the Internet, are part of the online world. (Local area networks allow communication in limited environments, such as inside an organization.) In other words, the Internet and the World Wide Web are part of the online communications world; they are not the entire online world. However, the Internet and the Web are among the most important parts of the online world for mass communication, because they are where much digital, online media content resides and is available to the public.

Digital media of course plays a role in every communication function that analog media plays, ranging from surveillance to entertainment. Whether democracy and society ultimately will be better served by an Internet-connected society is impossible to say. However, there is no doubt digitization and online media are changing and will continue to change mass communication and the public that receives almost all of its entertainment and information from mass media.

A 24/7 media environment is quickly emerging, if not already here. Newspapers have had to create policies so their online versions do not scoop their printed morning edition the next day-as much so they do not tip off their competitors as for not wanting to hurt their own newspaper sales. With cable TV there is of course 24-hour entertainment and news, and the always-on nature of the Internet has taught the public that they can obtain specific media content on demand as long as they are connected to the Internet.

Even media that we are accustomed to thinking of as "set" once created, such as magazines or books, can easily be updated at any time if they are distributed in digital form. A nonstop media environment, although increasing the opportunities for news, information, and entertainment, also has its negative side. With new messages being delivered all the time and a huge choice of media to interact with, it can be very easy to become distracted with flashy entertainment or essentially "drown" in information.

The pervasiveness of the media system means that wherever one goes, there is likely to be unprecedented access to mass communication. A new media pioneer who works as a civilian researcher in a naval research lab once remarked that he was leaving for a week's vacation in the Caribbean on an island where there was no Internet, no phone service, and no communication services of any kind. He had to escape. Unfortunately, with today's global satellite communications, it's not possible to truly "escape" anywhere on the planet.

Increasingly portable media devices and flat-panel screen technology improvements also mean that we have a growing ability to take our media with us and access it (or have it thrust upon us) in places where we previously did not encounter media. Displays in elevators are one example of how advertisers are using technology to reach a captive audience. Pervasive mass communication means better access to entertainment, commercial information, and news. Such access means there is at least the potential for a better-functioning democracy, because more information is available. Of course, better access may not come evenly to all or allow everyone to benefit equally from that access.

Simply providing more information into the media system may often result in a widening, rather than a narrowing, of the gap in knowledge between those in high and low socioeconomic groups.Personal Information Space The convergence of digital media is leading to the development of a personal information space. A personal information space is a virtual location assembled and accessed online where an individual keeps data, or information. It is more than just a digital personal library, however. In a personal information space, one can process private voice, fax, and e-mail communications, and create, access, and store Web-based media content, including multimedia, all from any location around the world.

No longer is one's personal information space limited by geography, time, or culture. One can access a personal information space when mobile computing and communications devices are connected to global, wired, and wireless telecommunications networks. Increasingly, the personal information space is being integrated with public information space (i.e., the content generated by media organizations and others). Unfortunately, the personal information space is also subject to various threats and dangers, such as erosion of privacy, computer hackers, or technical failures.

## Internet Technologies

Prior to the era of the Internet, institutions or organizations that had computers had no simple way for the computers to communicate with each other, even if they were connected by a wire, as computers ran machine-specific languages and programs that could not be understood by other computers. In 1969, the foundations for the Internet were laid when the Defense Advanced Research Projects Agency (DARPA) launched the Advanced Research Projects Agency network, or ARPAnet. ARPAnet was the first national computer network, connecting many universities around the country for advanced, high-speed computing applications and research.

It was not yet the Internet, but it was the beginning of online communications. But there was still no "common language," or protocol, that computers could use to easily transmit information via the network.

Although it is difficult to pin down an exact date when the Internet officially started, in 1982 the Defense Department adopted TCP/IP as the basis for the ARPAnet. Moreover, at this time researchers began defining an "internet" (lower case i) as a connected set of networks using TCP/IP, and the "Internet" as a set of connected TCP/IP internets.Creating the World Wide Web For the first decade of the Internet's existence, its usage was limited largely to researchers. Use of the Internet required knowledge of a variety of arcane commands and terminology. The limited, specialized nature of the Internet underwent a fundamental change in 1991 when Tim Berners-Lee, an MIT researcher at a physics laboratory in Switzerland, invented the World Wide Web and began to open the use of the Internet to a much wider set of users. The advent of the Web as a global publishing medium made possible the most fundamental shift in human communication since the advent of the printing press five centuries earlier. The Web enabled easy many-to-many communication over distance and time.

In addition, in contrast to traditional media of mass communication, anyone can create and publish on the Web for very little cost or expertise. The World Wide Web (WWW) is a subset of the Internet and is perhaps best described as a global electronic publishing medium accessed through the Internet. Technically speaking, the Web is made up of an interconnected set of computer servers on the Internet that subscribe to a set of TCP/IP network interface protocols. These technical protocols include assigning to a Web site a Uniform Resource Locator (URL) based on its TCP/IP Internet address, which is the Web site address that Web users are familiar with. URLs include the instructions that are read by a Web browser, a navigational tool to travel the Web.

A Web page is any document, or collection of content, that resides on a Web site. The content can take any form, including text, graphics, photographs, audio, video, or interactive features, such as surveys or discussion forums. A Web site can consist of one page of content or many such pages and can include hyperlinks to other Web sites. Content on a Web page is tagged, or marked up, using what is known as hypertext markup language, or HTML, to format the content so it displays correctly on a screen. In addition, each document uses hypertext transfer protocol (HTTP), which enables the standardized transfer of text, audio, and video files, as well as e-mail from one address to another.

Another huge gain in making the Internet accessible to even more people was the creation in 1993 of Mosaic by Marc Andreeson, then at the National Center for SuperComputing Applications (NCSA) at the University of Illinois at Champaign-Urbana. Mosaic, which eventually became Netscape, provided a graphical user interface with the Web that computer users who had Macs or Windows PCs could quickly understand and use. Although GUI browsers Viola and Erwise were also created in 1992, by the end of the year Mosaic was being written about in mainstream media and became the most well-known Web browser.

Microsoft created their own graphical browser, Internet Explorer (IE), in 1996 to compete with Netscape's browser, then called Netscape Navigator. By offering Internet Explorer free and eventually bundling it with the Windows operating system, IE was able to become the dominant Web browser in only four years, with 75 percent usage compared to Netscape's 25 percent. In 1999 Netscape was bought by AOL, a year before AOL acquired Time Warner. As any regular Web user can attest, Web sites and text size often look different not only on different browsers but on different versions of the same browser. More advanced functions or codes on Web sites may not show up on earlier browser versions, thus making it difficult for Web designers and content companies to create Web sites that are consistent across the online audience.

Reflecting the continued dramatic, almost exponential growth of the Web, by 2001 millions of individuals and organizations had published "home pages" on the Web. With the cost barriers to entry in the global publishing arena removed or dramatically lowered, the Web for the first time brought press freedom to virtually any citizen with access to a computer and a phone line. Of course, more than half the world lack either or both of these, so we're still a long way off from realizing the dream of fully democratic communications.

In some Scandinavian countries Internet penetration has reached 75 percent. So even traditional mass media aren't by any means necessarily universal relative to the Internet and the Web. Yet, when one considers the fact that only a tiny fraction of the U.S. or world's population has its own printing press, television, or radio station, the Web has dramatically increased the diversity of media voices available. Of course many of these voices are lost in the global cacophony of the Web, but that is a different issue.

In its broadest sense, the Internet encompasses virtually all other media, as well as a broad cross-section of human culture, commerce, and creation. Virtually anything one can think of almost certainly exists in some form on the Internet, from information in the Library of Congress to deliberately misleading information created by hate groups or individuals. As distinguished communications scholar Fred Williams once observed, "Going on the Internet is like going through someone else's trash."

The Internet as we know it now is not controlled by any one person or organization. It is a medium of multimodal (i.e., it involves the various senses) content and interactive communications, in particular e-mail, where users (and organizations) send and receive text and increasingly audio, video, photos, graphics, and hyperlinks. Although much of the content on the Internet is available free to users, there is a growing trend among media and technology companies to start charging consumers, especially after advertising dropped off sharply after the dot-com meltdown in 2000. Content or services that were previously free, such as a given amount of disk space or the ability to check remote e-mail accounts through Web e-mail accounts such as Hotmail, for example, were starting to be offered only to paying consumers. This trend of partially free services or content and payment for what is considered "premium" content can also be seen among media companies online, and it is still too early to tell whether these strategies will be successful in generating revenues.

Online media companies are also experimenting with increasingly obtrusive online advertising, such as larger pop-up ad windows or automatic pop-under ads that are changing the look and experience of the Web for users. In the early 1990s there were many debates among users of online discussion groups about whether any advertising should even be allowed on the Internet, considering the fact that it was created by using taxpayers' money.

## Computer-based Networks

Networks have long been a fundamental part of mass communication, especially broadcasting. Networks, or systems of interconnected communication vehicles, can take many forms. In broadcasting, networks of affiliated local stations provide a means of distributing shared programming to audiences across large geographic regions, most typically the entire country. Traditionally, these networks have been linked through analog technology. Today's networks are increasingly digital. This offers several advantages in mass communication, most importantly the ability to transmit multiple streams of information, including audio and video content, text, or other types of data. In addition, computer-based networks permit the distribution of content to be more flexible, so if one route of the network is unavailable, the network can automatically and efficiently reassemble through a different route.

Digital networks permit compression of data, which enables a mass communicator to send more information in the same amount of space on the network. Finally, a computer network easily permits downstream and upstream communications (from the source to the receiver, and vice versa), whereas traditional networks were largely one way and relatively expensive.

A modem (a foreshortening of modulate-demodulate) is used to convert the digital information in a computer to analog signals for transmission over a phone or cable TV line and to convert analog signals to digital information for a computer. Modems can also operate wirelessly by converting data into radio signals. Modems were invented in the 1960s, before the days of the personal computer, as a means of letting "dumb terminals" (i.e., electronic machines where users typed information on keyboards) dial into a remote computer.

These early modems converted the typed characters into audio tones that were sent over the telephone line and then were converted back on the other end of the line where they were received by the computer, which processed the information received. Today's modems operate in a much more efficient and faster fashion. Whereas early modems were hard-pressed to convert and transmit 300 bits per second, today's modems can convert and send millions of bits per second. This allows them to rapidly send or receive not just text but also audio and video.

Bandwidth is a crucial element for online communication to reach its full potential to be a mass medium. Without what is called high-speed, high-bandwidth, or broadband connections to the Internet, most people online are unable to receive audio or video in real time or at the same quality as they are used to from television or radio. Narrowband is the term used for low bandwidth communications, such as dial-up phone modems. Bandwidth available for Internet service has traditionally been narrowband via dial-up modems, typically delivering anywhere from 28 kilobits to 56 kilobits per second. Video at these narrowband rates is very limited, usually a small window of jerky motion, of low-resolution imagery and only marginally better sound. In a technical sense, bandwidth refers to the electromagnetic frequency or spectrum available for delivering content. Bell Labs scientist Claude E. Shannon in 1948 provided a precise mathematical definition of bandwidth, defining the capacity of a communications channel in terms of bits per second. A voice phone call, for example, uses about 3000 hertz (Hz) bandwidth, whereas a telephone modem operating at about 33.6 kilobits per second requires a little more bandwidth, or about 3200 Hz.

Think of bandwidth not so much as electromagnetic frequency, however, but more in terms of how large a pipe is that comes to your home delivering data rather than a physical thing like water. Someone who is able to tap the large "data pipe" directly can access the flow of data at equally high speeds. Sometimes this is called a "fat pipe." However, if the pipe that accesses the main pipes is very thin, data will come at only a trickle, no matter how fast his or her personal computer is. As of 2002, the Internet was relatively limited in terms of available bandwidth. It varied depending on how a user connected to the Internet, such as a telephone modem or a cable modem, although a second generation of the Internet known as Internet 2 brought 45,000 times more bandwidth. At that time the Internet 2 was already providing a very high-speed connection among 170 universities. It promises in the future to bring such high-speed connectivity to a broader portion of society.

One of the great challenges that cable and telephone providers have begun to solve in the past decade is the so-called "last mile" obstacle (not always literally the last mile, but somewhere from a few hundred yards to more than a mile or so). For more than a decade, many cable and telephone companies have had considerable bandwidth available in their backbone or trunk lines but have run into a problem connecting these broadband facilities over the "last mile" to the end consumer. The cost of making these final connections has been prohibitive. But as the cost of the technology has fallen and various technological innovations have occurred, it has become increasingly feasible to provide the last-mile connections. When this is completed, it means that consumers will have the same broadband access that currently only large organizations have.

## Telecommunication Technologies

In the world of telephony, development of the digital subscriber line (DSL) has made it possible to use standard telephone wire, the twisted-pair copper wire in most homes, to provide relatively low-cost broadband capabilities in the home. Other emerging network technologies, such as aDSL (asymmetric DSL, which means high-speed downstream, much slower upstream), are also being deployed by the phone companies. Telephone companies have lagged somewhat behind in their delivery of highspeed Internet and broadband digital video services to the home, with DSL available in only a handful of markets.

Many customers have complained bitterly about late or no installation, technical troubles with little customer support, and lack of communication between the telephone companies and the Internet Service Providers (ISPs) that have hampered easy installation of DSL services. In the cable TV world, the cable modem and the set-top box have made it practical for many cable systems to begin rolling out digital cable services that not only provide digital TV program services to the home but also permit highspeed Internet access and telephone service. Broadband content delivery via already existing coaxial cable systems mean that cable modems are roughly 174 times faster than 56k modems. One problem of the cable modem is that the network is not switched; the network bandwidth is shared by the users within each local geographic area, or node. In other words, the more users on a given cable modem node, the slower the network becomes. Cable companies can put fewer users on nodes by increasing the number of nodes, but this of course increases their costs. TCI has already indicated that movies on demand or video on demand will be limited in its cable modem service to about 15 minutes in duration.

Bandwidth availability is gradually increasing as low-cost bandwidth rolls out nationwide in the first years of the twenty-first century and low-cost digital consumer access devices enter the marketplace. The importance of broadband capability to usage patterns on the Internet may be profound.Research shows that Internet users with broadband access already have substantially different behaviors than when they used dial-up connections. One trend that has been noted is that they are more likely to create and distribute media content than dial-up users. Online expenditures more than double for users of broadband services.

David Clark says that broadband Internet could bring "Real-time high-fidelity music, telephone, video-conferencing, television and radio programs.... There will be new entertainment options, such as movies-on-demand, and new features, such as the ability to call up information about a movie's director or its actors as they appear on screen. Users will be able to play online games-live-against many contestants scattered around the globe."Of course, the data on high-bandwidth users are hard to interpret, because those who have access to broadband capabilities in 2002 are, in general, what are called "early adopters." They tend to have more income, are more educated, and are likely to be two-parent households. They are different by nature than the late adopters, and it may be in part these differences that account for the differences in how they connect to the Internet. In a sense, talking about online mass communication means talking about all of mass communication. Today, countless offline mass communication products, whether print or electronic, also exist in some form online, whether for free or fee. The American Journalism Review counts 4925 online newspaper sites, and all of the top-50 magazines by circulation have Web sites, as do thousands of specialized periodicals. Some use their Web sites mainly to showcase their printed material, but others publish original material online as well. As bandwidth increases, most television stations will place more of their content online, just as an increasing number of radio stations do already. Similarly, more of the cinema and recording arts will become available online as connectivity and bandwidth continue to increase. Many mass communication sites originally created solely online are also available, from one-person operations to major endeavors by large companies such as Microsoft in their creation of the online magazine Slate. Even companies not traditionally involved with mass communication have found that maintaining their own Web sites is an excellent way to communicate directly to a large audience.

## Digital Newspapers

Newspapers are among the leading providers of online news. Online newspaper sites also face direct competition on a number of other fronts in the online arena, including to some degree the general public, who can put online eyewitness accounts of events. Matt Drudge's site was ranked twentieth in August 2002 of top news sites. News services such as AP or Reuters offer their breaking news content online on news aggregator/portal sites such as Yahoo! News, which was in third place. Newspaper Web sites have the potential to take back some of the ground in breaking news coverage that they lost to television, but this can create an odd situation of a newspaper Web site "scooping" its more established print version, which comes out on a set schedule.

The competition online newspapers face with cable and broadcast television news channels shows in part the complexity of the online mass communication world. For example, there is nothing stopping a newspaper from training some of their journalists to shoot and edit video footage, which could then be webcast on the newspaper's site. The same could apply to audio recording. This is exactly what some newspapers have been doing, such as the Tribune Company, whose flagship newspaper the Chicago Tribune and flagship television station WGN are part of a converged news operation providing considerable quality online video news.

Advertising is another big area in which newspaper companies are threatened by the Internet, especially classified ads. Classified ads make up a large portion of advertising revenues for most newspapers, but online classified ads aggregators can allow people to search far more ads than are usually covered in a newspaper's circulation area. If a consumer wants to look for an item in a region that is covered by four or five different newspapers, there is no reason for that person to go to each newspaper's Web site to look at classifieds when he or she can see them all on one Web site. Likewise, a consumer can visit an online auction site like eBay and reach a nationwide group of potential sellers or buyers.

## Digital Books

Digital books offer a variety of advantages over printed books, including permitting the reader not only to read the text but to make electronic annotations and bookmarks, access the content of the book via an interactive table of contents, and keyword search the entire text. Digital books and their sale online also point to a fundamental issue in the future of how mass communication industries will derive much if not most of their revenue: from online transactions, or what is called e-commerce. Although few people still may want to read a book on a standard computer screen, the Internet has brought a growing number of books online and made them available universally. Much of the content of these online books may be most valuable as a research tool, but many may also find them a potentially viable way to read for pleasure. Segments of a book can be printed out, for example, saving weight and space when traveling, and as computer screen technology gets better and creates sharper images, reading on-screen will not be as tiring as it can be today.

In the late 1990s, major publishers prepared for a growing surge in consumer demand for electronic books, and many experimented with the online sale and distribution of digital books. However, after the economic slowdown in 2000, which hit technology and Internet companies particularly hard, and a subsequent slowdown in sales of digital books, publishers started to adopt a slower, more cautious approach. However, many believe that just as young people powered the paperback revolution, the young are showing a voracious appetite for digital books. Dennis Dillon, a librarian at the University of Texas, was initially surprised at the popularity of electronic books.

"No one was sure whether anyone was going to read any digital books," said Dillon, the university's head of collections and information resources. "We were somewhat skeptical." After minimal promotion of the University of Texas's newly purchased electronic books, with titles from Euthanasia: A Reference Handbook to From Barbie to Mortal Kombat: Gender and Computer Games, they are suddenly the most popular titles in the library. "Usually a book has a one-third chance of being checked out. So to have some titles checked out 25 times in two months-that's shocking."Some of the leading online book publishers and distributors conduct their business exclusively online. Among the leaders are netLibrary, peanutpress, 1stBooks and Xlibris. NetLibrary and peanutpress sell and distribute exclusively digital books online. They carry large numbers of titles, ranging from books published by university presses to trade books. Xlibris and 1stBooks are print-on-demand (POD) publishers who specialize in helping authors publish their own books, but they also offer digital books for online distribution.

## Digital Magazines

Despite the low bandwidth requirements for online text, online magazines have not been very successful to date. There are very few purely online magazines, Slate.com and Salon.com being two of the most prominent ones, but even these two online magazines have not been profitable despite publishing articles from world-class writers. Print-based magazines have done little original online content, preferring to use online versions of their magazines mainly for promoting print stories or archiving past issues. There are several likely reasons that online magazines have fared poorly, despite the apparent benefits the online world would seem to bring to magazines and long-form nonfiction writing in particular because of the lack of space limitations.

One of the main reasons is technological; reading long blocks of text on computer monitors becomes tiresome because the resolution of computer monitors has not reached the same quality as text on a printed page. Even once computer monitors do achieve higher resolutions, there is still the issue of portability. In other words, unless monitors can compete with a printed magazine in terms of clarity and portability (and to some extent disposability), there is likely to be a barrier among people in reading anything of any length in a digital format. This of course is not unique to magazines, as digital books and newspapers face the same issues. E-ink or some other digital form of paper may solve this problem, but this will not reach consumers for a few years yet. Changing behavior among the public in how they consume media is another factor weighing against long-form, text-based narratives like those found in The New Yorker or The Atlantic Monthly.

With greater competition for their attention during the day, the public seems less willing to spend the time necessary to read long articles. In an online environment, readers are also more likely to click on hyperlinks to get more information rather than wait for the author to fully explain a point. Likewise, an article may be a starting point for a much more robust discussion among Web users in the form of online discussion groups. However, there could be an important niche for magazines to provide a measure of distance and perspective on issues and to thereby be a voice of authority in their specialized area. If a magazine is perceived as having suitable authority and is respected highly enough by its readers, it seems that the public is willing to pay for online subscriptions. Consumer Reports Online is one of the few success stories in the online magazine world, with over 580,000 subscribers who can access archived articles and reviews on various products.

## Digital Radio

Thousands of radio stations offer live webcasts of their regular over-the-air transmissions, although there are unsettled issues regarding royalty payments for webcasts of recorded music. In 2002, the U.S. Copyright Office decided that webcasters must pay 70 cents per song heard by 1000 listeners, half of what a government panel recommended they pay in a decision in February 2000 but still less than what the recording industry was asking. Webcasters also have to pay royalties retroactively from 1998, and many of them said the rates and retroactive payments would cost them hundreds of thousands of dollars and force them to shut down. Traditional radio stations are exempt from this new royalty payment method, successfully arguing in 1998 that they are helping promote the music. The term radio is itself somewhat meaningless in an online digital world in which music is not broadcast over the air but retrieved by users on demand over the Internet. In that way, online "radio" is technically audio programming, and in some ways will seem much closer to the current state of on-demand online music downloads, even though the term radio may stay with us or change its meaning.

Regardless of the technical approach, online-only radio stations such as spinner. com are free of many FCC regulations and don't require a license to operate. Moreover, they reach a global audience, and there's no potential for signal interference. As a result, there are hundreds, possibly thousands of online radio stations. Many of these stations operate independently of any broadcast parent, although legislation is being drafted that will bring them under some of the same rules broadcasters face.

Online radio stations must abide by the Digital Millennium Copyright Act, however, which prohibits the advance posting of playlists, which would permit users to plan in advance when to copy songs. Even before the 2002 rulings on royalty payments for webcasts, some online radio stations paid royalties to one of the two music licensers in the United States, the American Society of Composers, Authors and Publishers (ASCAP), or Broadcast Music Inc. (BMI). Live transmissions of sports events have emerged as an important part of online radio. In the case of major league baseball, Internet users must pay a fee to listen to live audio webcasts of professional baseball games. Some other professional sports, such as the National Football League (NFL), make the audiocasts of their games available for free online. One of the challenges of listening to online radio is finding stations that are available. One list with more than 10,000 Internet radio stations from all around the world was created by WMBR-FM at the Massachusetts Institute of Technology. It allows for searches by country or region, call letters, format, or frequency.

## Digital Music

The MP3 file format represents an interesting example to illustrate how the new networked digital marketplace works. It's indicative how the recording industry is likely to sell much of its music in the future. The $5 billion recording industry is faced with a dramatic shift as users embrace digital audio and all the flexibility and power it affords them. The MP3 file format is used to compress near-CD-quality audio of music, which then lets the files, which generally range between 2 and 4 MB in size, be distributed easily over the Internet. There is nothing inherently illegal in the MP3 file format itself; it is simply the technology used to compress audio files. The problems with the recording industry arise because consumers can copy, download, and distribute music largely for free through file-swapping services such as KaZaA. The recording industry has long held that it is acceptable for an individual who purchases an album, cassette, or CD to make a copy for personal use, and it is of course not uncommon for friends to exchange copies of music among themselves.

File-sharing of online music takes these exchanges to completely new levels, how-ever. At any given time during the day, it is not unusual for 1.8 million users to be on a service such as KaZaA, exchanging close to 300 million files. These are not simply audio, of course, and can include video, text, and images as well, but it would be safe to say most of the files being shared are music files. KaZaA, Morpheus, Grokster, the now-defunct Napster, and many other online peer-to-peer file sharing systems have been at the core of this phenomenon. For now, suffice it to say that the major labels and everyone else in the recording industry value chain are enormously threatened by this development. They lose significant control and, they say, their profits and investments are being reduced along with the piracy of their intellectual property.

In 2001, the major record labels finally began getting into the online music business themselves even as they maintained their lawsuits against file-swapping services. Warner Brothers, BMG, EMI, Arista, Virgin, Jive/ Zomba, and others teamed up to launch MusicNet.com, a for-profit online music distribution service. Consumers pay $9.95 a month to RealOne, which uses Real Networks to enable music downloads. Even as the labels become more involved in digital distribution of their music-belatedly, many would say-other industries within the current music distribution industry will also be faced with drastic changes, such as those in the plastics industry, distributors, and retailers.

Finding new songs and artists that you may like may become more difficult than ever in the digital age. As production and distribution costs decrease, more people can create their own professional-sounding (although not necessarily good) recordings and distribute them digitally. One solution that is emerging is the use of intelligent agent-based collaborative filtering, a sort of music club for the digital age. Amazon.com and other e-commerce organizations already are making use of collaborative filtering to alert their customers to new books and music that they would be likely to find appealing, based on previous purchases and the purchasing patterns of people with similar buying tastes.

## Digital Film

The Internet has proven a valuable means of distribution for film, especially independent and short films, which have traditionally found distribution to be the most vexing bottleneck in becoming successful. Because independent and short films vary widely in quality and style and are produced outside the mainstream of Hollywood, few theaters, or theater chains, have been interested in showing them. Because of the low cost of online distribution and the potential to build audiences over time and space, a growing number of Web sites provide extensive independent and short films online, typically for free viewing. Two of the leading independent film sites are Indie Films and ifilm. These advertising and e-commerce-supported sites feature hundreds of independent films and film trailers for free viewing. BMW Films is another interesting example of an online film site, in this case funding the production of independent short films-as long as all funded and shown films feature BMW cars. As commercially oriented as this may sound, the films produced have been of very high quality and not too crass in their commercialism. The quality of the online films available on the site, some done by famous directors, have been praised widely by leading independent critics.

## Digital Television

There is a wide variety of sources of video content online. Most television stations and all major networks maintain Web sites. A small but growing number of them provide at least some of their programming via the Web.

An even smaller but growing number of these and other programmers provide near-broadcastquality video programming via the Web, some of it on demand. Viewers can go to the station's Web site and select from a variety of news and information options, including stories reported in text, audio, or video format. These various formats help add depth to the stories that video-only formats simply are not able to do because of the need for interesting visuals, sometimes at the expense of exploring related issues to a story.

For established broadcast or cable stations, changing to a digital format that allows them to take full advantage of the capabilities of online video can be expensive. Jim Topping, former general manager of KGO-TV (KRON's ABC-owned competition in San Francisco) and then senior vice president (now retired) at ABC-owned TV stations, notes that the installation of a powerful video server cost KGO-TV $9 million. Other technology to help the station become digital cost another $12 to 14 million. This investment gives KGO-TV the capability of delivering any of its programming at high quality and on-demand via the Internet or other broadband media. However, the investment KGO-TV made is prohibitive for the majority of stations, especially those in smaller markets.

Of course, a $9 million investment is not necessary to start transmitting digital video on the Internet. One small-market station has found a way around the high price tag. In Kingsport, Tennessee, DMA 93, WKPT got a digital signal on air for just $125,000. George DeVault, president of Holston Valley Broadcasting Corp., owner of the ABC-affiliated station, decided to forego HDTV initially and bought and installed a digital system to put a standard definition television signal on air. This low-cost approach is what has opened the door to many independent producers who are now transmitting television-like programming on the Internet. Public television stations are also converting to digital television.

A June 1999 Arbitron NewMedia Internet study showed that at that time almost three-quarters of Internet users in the United States spent up to 30 minutes a week watching streaming video. Nearly half planned to watch more streaming video in the future. The development of television programming on the Internet has given rise to at least two main types of online video programming: that which has been transferred from off-line television and original Internet video programming. Because so much of online video or television is still in its infancy and still largely constrained by technological issues such as bandwidth, computer storage capacity, and screen resolution, it is likely that these two types will morph or change in ways that we cannot foresee and create some kind of as yet uninvented form of online programming.

An early example of the type of hybrid that may develop is Alternative Entertainment Network TV (AENTV), which combines original Internet video programming and aggregates digital video originally aired on television. AENTV was named one of the "10 Great Video Sites on the Internet" by Broadcasting & Cable magazine and has collected hundreds of hours of programming from a variety of TV shows, all available on-demand and for free. This type of programming, which simply serves up programming originally produced for television, is still perhaps most common, although there is no definitive study yet available to prove this. Although there are examples from virtually every type of programming category, most "television" programming of this type is simply promotional.

Trailers, soap opera clips, and sit-com excerpts are put online to generate viewer interest in an upcoming show that will appear on television. One exception is Comedy Central, which provides much of its television programming online on-demand. The most common type of nonpromotional programming thus far from overthe-air television stations is news and public affairs. CNN.com, MSNBC.com, CBS News, ABC News, Fox News, and NBC News are among the leaders in providing online video news. The CBS Boston affiliate WBZ provides exceptional regional online video news, as do Seattle CBS affiliate KIRO and NBC San Francisco affiliate KRON.

A leading provider of international video news online is the BBC World, which provides live video news feeds online. Viacom's 2000 acquisition of CBS has begun opening up greater opportunities for television, or video, on the Internet. With its new Internet division, MTVi, including both MTV and VH1, as well as its recently acquired SonicNet, Viacom CBS is positioned well to provide not only online music but also online music video. Combined with CBS video news strength, the Viacom CBS empire is poised to take a leadership position on many online television fronts. DTV can be interactive, which opens the door to more participation and interest in shows among viewers.

ABC's Enhanced TV synchronizes customized interactive content over the Web during many of its programs, such as Who Wants to be a Millionaire and Monday Night Football, which offers an interactive play-along synchronized game as well as live statistics and facts about the players and teams. ABC has initially focused on the "two-screen" platform whereby viewers have an Internet-connected computer in the same room as their television, so it's not truly interactive television. There are more than 40 million homes that today are capable of such TV-Internet convergence, which is closer to a critical mass than the current set-top box installation numbers. Currently several companies are trying to make the convergence complete, either as a PC-turned-TV or a TV-turned-PC.

Television programming typically has been provided to the public by a limited set of program providers. They have made their programming available on a scheduled basis, packaged for mass audiences and broadcast according to a controlled schedule. Just as online music distribution opens the music recording industry to potentially thousands of new artists (i.e., program providers), the Internet opens the TV business to thousands, perhaps millions, of new TV program providers. Internet-original programming began with web cams, tiny cameras that attach to a computer that have enjoyed enormous popularity around the world.

Web cams have provided typically live feeds of everything from a coffee pot at Oxford University to a seemingly endless series of individuals inviting the public to pay to watch their uncensored private lives. This early experimentation among amateurs has given rise to much more serious original online video programming, however. Some of the most interesting are web cams set up by researchers to allow viewers from around the world to observe important, unusual, or simply interesting experiments or other research in action. Universities and not-for-profit organizations have also used online video to show lectures from famous guest speakers, panel discussions, or to highlight on-the-scene video footage shot elsewhere in the world that does not make it to mainstream media.

Online programming guides, also known as electronic program guides (EPGs) in digital television or video, are becoming a necessity as Liberty Media chairman John Malone's outdated notion of a "500-channel universe" is replaced by a "million-channel universe." Leading the way in the online program guides today is tvguide.com, the online version of TV Guide, with 50 million subscribers. Tvguide.com offers viewers a fully interactive and keyword-searchable guide to the coming week's television programming from terrestrial broadcast to cable, satellite, and the Internet. The site also features daily news about television and other media, a database on more than 40,000 movies, and digital video.

The same developments that have challenged the recording industry will start to emerge in the television industry as digital television enters the marketplace and broadband use increases, allowing for fast downloads of large files. Digital video files are much larger than what most people can accommodate in downloads, which has hampered the popularity of video file-swapping online. Fixed media, such as videotapes or DVDs, are still primarily used for on-demand, repeated viewing, although new technologies could easily change this as well and make fixed media obsolete.

The television industry has been closely watching what the recording industry is doing with the hope that they will not repeat the same mistakes. However, concerns about intellectual property theft and unwanted competition have limited some programmers' forays online. Large television companies, broadcasters, and cable companies generally see the Internet and digital video as threatening their loss of control and their existing franchises, and whether they learn from the experiences of the recording industry or follow them in litigation and political lobbying to stop potential threats to their control remains to be seen. So far, attempts at creating secure digital media have not been successful. They have either been extremely clumsy and overly restricted consumers' ability to play media such as a CD on a variety of devices, have been foiled by low-tech mechanisms such as using a black marker pen to mask digital encoding, or have been extremely unpopular with consumers and thus failed, such as what hap- pened with the DIVX format that forced consumers to pay per use of a digital video they "bought."

Traditionally, media enterprises, like other industrial-age businesses, required heavy capital investment. Capital was required to build the transmission towers for television and radio stations. It was required to buy printing presses and to build cable systems. Not only was the infrastructure expensive, but it created enormous barriers to entry for prospective competition. In the digital age, these rules are changing. Some of the most significant barriers to entry are shrinking. Capital costs are reduced, as anyone can go online and create a Web site and potentially compete with established media companies. It's part of why relative newcomers such as AOL, Yahoo!, and eBay have been so successful. Of course, the cost of producing quality original content is still relatively high, especially for video but less so for text or audio. Moreover, few media enterprises have found a formula for making profits in the online world, as most people seem unwilling to pay for content and advertising does not seem to be very effective. Most large online media outlets have the backing of traditional media companies behind them, such as MSNBC.com, a joint venture between NBC and Microsoft, or CNBC.com. With advertising and other revenues being generated from the parent companies, these online endeavors can afford to take losses for a while as online revenue models can be tried and tested. Many smaller, independent media dot-com companies were not so fortunate, especially after the economy and stock market spiralled downward from spring 2000. Some severely scaled back their business plans, and others, such as award-winning online crime news Web site APBNews.com, went out of business. Other online media enterprises such as Salon.com, which are still creating quality original content, are now charging for what used to be free.

## Digital Libraries

Every library is different, every digital library is different, and different players are advancing many definitions for the digital library. Arms, for instance, defines a digital library as: a managed collection of information, with associated services, where the information is stored in digital formats and accessible over a network. A crucial part of this definition is that the information is managed. For the Digital Library Federation in the USA: Digital libraries are organizations that provide the resources, including the specialized staff, to select, structure, offer intellectual access to, interpret, distribute, preserve the integrity of, and ensure the persistence over time of collections of digital works so that they are readily and economically available for use by a defined community or set of communities.

Some prefer to work with the concept of the 'hybrid' library, as they believe that this more closely describes the reality of libraries, which have always been hybrid. The digital is one more (albeit very different) format that librarians have to deal with in a multi-format environment, where for many years non-documentary mixed-media objects, both analogue and digital, have been growing in importance, and where technology for access has had to be provided. Most libraries can supply microfilm and microfiche readers, video players, audio tape and CD players, as well as terminals for access to digital resources. However, there are often strategic reasons for defining activities as coming under a digital libraries heading - the possibility of securing funding, for instance, from programmes which are defined as digital library initiatives.

Hybrid libraries are designed to bring a range of technologies from different sources together in the context of a working library, and also to begin to explore integrated systems and services in both the electronic and print environments. They exist 'on the continuum between the conventional and digital library, where electronic and paper-based information sources are used alongside each other'. The concepts of virtual libraries or libraries without walls reflect the integrative possibilities inherent in the digital: if significant library collections are digital, then the confines of space no longer define boundaries upon information.

Virtual collections from many different sources can be assembled and accessed from anywhere, without the user even knowing where the sources reside, and personal virtual collections can be built to serve many purposes. Hence a virtual library could potentially be enormous, linking huge collections from all around the world together, or it could be very small, being the personal digital collection of one individual. Digital libraries are, like any so-called revolutionary change, a development of a whole range of underlying theories and technologies that have come together to create a paradigm shift.

The speed of recent developments has taken some librarians by surprise, especially the exponential growth of the amount of digital data available, but they are understandable if we look at some of the precursors that have led to the current trends: new developments rarely spring fully formed from the ether. There are many different kinds of digital libraries creating, delivering and preserving digital objects that derive from many different formats of underlying data, and it is very difficult to formulate a definition that encapsulates all these. Some principles are:

- A digital library is a managed collection of digital objects.
- The digital objects are created or collected according to principles of collection development.
- The digital objects are made available in a cohesive manner, supported by services necessary to allow users to retrieve and exploit the resources just as they would any other library materials.

— The digital objects are treated as long-term stable resources and appropriate processes are applied to them to ensure their quality and survivability.

Vannevar Bush, one of Roosevelt's advisers in World War II is generally credited with being the first thinker to suggest mechanical and electronic means of dealing with complex information, and of finding paths through information universes that could be marked for others to follow. Bush was grappling with the problem that has beset humanity since the invention of printing, and possibly before. How, when so much information is available, does one remember what has been read, make connections between facts and ideas, and store and retrieve personal data? Pre-literate humanity was capable of prodigious feats of memory, but the total amount of knowledge it was necessary to absorb was orders of magnitude smaller than it is now. When writing became widespread, personal notebooks and handbooks were used as aides-memoires, but these could handle only a fraction of the data that even one individual had to cope with by the 1940s.

Emanuel Goldberg developed very high-resolution microfilm, and also the technology underlying microdots, in the 1920s. In 1932, he wrote a paper describing the design of a microfilm selector using a photoelectric cell, which Buckland sees as 'the first paper on electronic document retrieval'. Paul Otlet saw huge possibilities in new forms of information recording and transmission: he was aware of the information and communication possibilities of telegraph, telephone, radio, television, cinema and sound recordings.

Paul Otlet visualized new kinds of work desks where there would be no documents, only a screen and a telephone with access to documents on film which could be called up at will. Otlet also formulated some new organizational principles which would hold all the documents in a linked relationship: a Universal Network for Information and Documentation which would connect centres of production, distribution and use regardless of subject matter or place - a development that uncannily presages the world wide web.

Though Otlet and Goldberg may have formulated the underlying theories upon which hypertext and information retrieval, and therefore digital library developments, came to be based, it was Bush's legacy that inspired later thinkers and developers. In 1962 Douglas Engelbart started work on the Augment project, which aimed to produce tools to aid human capabilities and productivity. He also was concerned that the information explosion meant that workers were having problems dealing with all the knowledge they needed to perform even relatively simple tasks, and Augment aimed to increase human capacity by the sharing of knowledge and information. His NLS (oN-Line System) allowed researchers on the project access to all stored working papers in a shared 'journal', which eventually had over 100,000 items in it, and was one of the largest early digital library systems. Engelbart is also credited with the invention of pointing devices, in particular the mouse in 1968.

The mid-1960s saw other pioneers conceiving of grand schemes that at the time seemed so innovative as to be impossible, but now are coming to realization. Ted Nelson designed his Xanadu system in 1965, in which all the books in all the world would be 'deeply intertwingled' (in his words - Nelson incidentally coined the word hypertext). Nelson also tackled the problems of copyrights and payments by proposing that there should be electronic copyright management systems that would keep track of what everyone everywhere was accessing, and charge accordingly through micro-payments. Impossible to implement at the time, these ideas are now commonplace, but it took Tim Berners-Lee to put them all together in the late 1980s.

## Creation of Digital Data

Despite the difficulties of the early technology, some scholars recognized the value of the computational manipulation of textual materials immediately. Text had to be painstakingly entered on punched cards or tape, but the benefits appeared so great in terms of retrieval and analysis that they persisted. Father Roberto Busa, for instance, formulated the idea of automated linguistic analysis of text in the years 1942-6, and started working with IBM in New York in 1947. He produced more than six million punched cards for his edition of the works of Thomas Aquinas, and in 1992 the first edition of his CD-ROM was published.

In the 1980s a new technology appeared in the form of the Kurzweil Data Entry Machine (KDEM), which was developed to provide texts for the blind. It used optical character recognition (OCR) to create electronic text for printing on Braille printers or (eventually) conversion to sound using text-to-speech processing. The KDEM was extremely expensive (in the tens of thousands of dollars), but it could be trained to recognize different typefaces and fonts, and it was relatively accurate - probably as accurate as modern OCR packages. Other text conversion needs could, of course, be satisfied using the KDEM, and large corpora of digital text were created for many purposes.

The Oxford Text Archive, for instance, gathered many of the texts that it still supplies as output from the KDEM service operated nationally by Oxford University until the early 1990s, and these have been used all over the world as the basis for digital text collections, at the Universities of Michigan and Virginia in the USA, for instance. Digital text is now everywhere, as output from word-processors, publishers and OCR engines. Interestingly, OCR is faster and cheaper than it was in the 1980s, but it is still not as accurate as it needs to be to replace rekeying and typesetting, especially if the materials are pre-20th century. Many library projects use OCR to capture bodies of printed text; others, where the text is too difficult or a higher level of accuracy is needed, use rekeying. However, there are some new developments in the processing of text that greatly improve retrieval from inaccurate OCR. For maximum usefulness, digital text needs more than representations of alphanumeric symbols on the printed page; it also needs metadata to record other information about the textual object from which it derived. Markup languages such as the Standard Generalized Markup Language (SGML), define textual metadata and specify complex schemas of standard metadata tags to inform all the processes to which digital text might be subject: description, retrieval, preservation, print output, and so on.

While early document-processing technologies generally operated on strings of linear text, the possibilities of hypertextual linking within documents were recognized too, though this was initially difficult to implement on a computer. Despite the claims of many modern gurus of cybernetic hypertext theory, written text is not linear but can be interlinked, interwoven and annotated on the printed page in highly complex structures not susceptible of computational representation by the first generations of hardware and software. From the mid-1980s, better screen technology, more processing power, the development of the mouse and the advent of the Graphical User Interface (GUI) have transformed the ability of the computer to represent and link documentary objects.

In the last five years, digital camera technology has developed to the extent that digital images can be captured that equal or even exceed largeformat analogue photographic reproductions. These have disadvantages, however: the cameras are expensive, and the file sizes huge, which causes problems for storage and delivery. However, from valuable originals, large archive images can be captured and stored for the long term, with lowerquality derivatives being delivered for viewing and printing. This is acceptable for most uses, and means that good-quality images can be integrated with other media for a more complete user experience.

As of 2001, the British Library has produced a high-quality digital facsimile of the 15th-century Sherborne Missal that is on display in its galleries on the largest touch screen in the UK. The unique feature of this resource is that the pages can be turned by hand, and it is possible to zoom in at any point on the page at the touch of a finger on the screen. High-quality sound reproduction accompanies the images, allowing users to hear the religious offices that make up the text being sung by a monastic choir. Now documents can be represented by high-quality images, with underlying searchable text, and with annotations in text, sound or video.

Cameras are being developed that can capture 3-D images, and there are sophisticated capture devices for digital sound and video, deriving in part from the huge growth in these formats in the entertainment industry. It is, of course, possible to treat the printed page as an image. Capturing digital images from printed pages has the advantage that the page will appear on the screen exactly as it was in the original. The disadvantage of this presentation is that the search and retrieval functions are lost, but a combination of both image and text representation means that documents can be represented by images, with underlying searchable text and with annotations.

There is a rhetoric that suggests that we are moving rapidly from print to digital media: this is not borne out by evidence from the publishing industry and from copyright libraries. The death of the book has been predicted with every new communication or entertainment technology, the telegraph, recorded speech, film, television and the internet, but the book seems to be thriving. At the beginning of the 21st century, the UK copyright libraries, for instance, are receiving more print material than ever before (in 1999 the figure was around 105,000 items for the British Library alone), and this is without the materials that they purchase.

Production of digital data is certainly on the increase, but it does not seem to be accompanied by a concomitant diminution in the printed output, though it is changing both the publishing industry and the world of libraries. In the 15th century, the introduction of the printing press industrialized the production of books, but did not do away with handwriting. What it did do was change book production from a cottage industry carried out in monasteries to a commercial industry, and incidentally resulted in a huge loss of authority on the part of the Christian Church in Europe.

Johannes Gutenberg, born sometime in the last decade of the 14th century, and trained as a goldsmith, produced the first book in Europe printed from movable cast type, the so-called 42-line Bible. Despite popular misconceptions, Gutenberg did not invent this method of printing, which was known in China and Korea from around the 11th century. What he did was to combine the technology of the goldsmith's punch with that of the winepress. The result was the printing press - a machine that combined flexibility, rapidity, and economy to allow the production of books that the increasingly literate, increasingly numerous European city dwellers could afford to buy and read. The 42-line Bible was produced in around 1455.

Hand-set movable type not only produced books faster than handwriting, it produced them in multiple copies. While with early printed books it is not quite true to say that every copy is the same, the differences are small compared with the differences between manuscript copies. And printed books can be produced in the hundreds and thousands, even millions, which is the significant advance. Evidence for the instant huge potential market for multiple copies can be found, according to Kilgour, in the estimates of the numbers of books printed in the last third of the 15th century, that is in the 30 years after the death of Gutenberg in 1568. These estimates suggest that there were some 12 to 20 million books printed - more than the number of all manuscripts produced in medieval Europe up to that time.

More books call for more readers, and so literacy rates rose rapidly in the next three centuries. Paper was made from linen rag during this period, but in the Napoleonic Wars rag was needed to make bandages, so wood-pulp paper processes were developed. Wood was plentiful and the process cheap, which meant that print output could increase. This happened in several ways: more publications were printed, and works were often longer, hence the growth in the multi-volume work of fiction in the 19th century. Periodical publications were produced which appeared monthly, then weekly, then daily and the mass media were born.

Early printing presses were wooden and were hand-operated, and the type was hand-set. This remained the case until the beginning of the 19th century, when things changed rapidly with one of those 'paradigm shifts'. Printing-press technology moved to iron presses, operated by steam, to mechanical type, to hotmetal type to phototypesetting and then to computer typesetting by the 1970s. Publishing began to become a separate industry from printing in the 18th century, and grew rapidly throughout the 19th century with the increasing mechanization of the processes, and the further growth of literacy rates and the demands of a reading public. 'The total book production of the nineteenth century exceeded that of the eighteenth by 440%', and many of the publishing giants that still control the industry (though in very different configurations) came into being during this period.

Computer typesetting brought an unanticipated by-product: electronic text. This was not always kept initially, and when it was, the typesetting codes used for marking up the text rendered it largely unusable for any other process. However, standardization of markup languages has meant that it is now possible to use electronic text for purposes other than printing, and many publishers have embraced electronic publishing in the last five years, especially in the production of journals.

With the revolution in printing and the birth of the publishing industry came new developments in libraries. In the Middle Ages in Europe, libraries were either monastic institutions holding relatively few, very valuable books or belonged to private individuals. Many of the great libraries of today are based upon some of the collections of these monasteries and individuals. When the British army invaded the city of Washington in 1814 and burned the Capitol, including the 3000-volume Library of Congress, Thomas Jefferson sold his personal library to the Congress to 'recommence' its library. The purchase of Jefferson's 6487 volumes for $23,940 was approved in 1815. In the 18th century, circulation libraries and subscription libraries started to appear to satisfy the demands of the larger reading public for cheap access to print, and the 19th century saw the rise of the great public libraries.

Librarians have always sought to mechanize routine processes as much as possible, and so were early adopters of computer technologies. Database programs were developed early in the history of computing for use in stock control, payroll systems and other commercial activities. The adoption of administrative systems for catalogue record creation was an early example of the use of computerized processes in libraries. This started as merely a means of producing, for manual filing, printed catalogue cards or slips, which would be referred to by library staff and users as a means of finding library resources. To extend the duplication of catalogues and thus the number of user access points to the catalogue, microformats were used, 'enabling multiple copies of the catalogue to be distributed for the first time and new sequences, such as title catalogues, to be introduced'.

Alongside this new capability to share catalogues, there started co-operative efforts to share the cataloguing load and to benefit from aggregating cataloguing effort across multiple libraries. Because bibliographic description is such a highly structured construct, the computerization of the catalogue was a significant and inevitable next step from these early mechanization initiatives. This was not the computerized catalogue as we know it now in the guise of the OPAC (Online Public Access Catalogue), but a straightforward listing of the library's resources with no links to borrower records or to external resources.

In tandem with the development of the computerized catalogue came the move to automate circulation functions. Initially, this was no more than associating a borrower number with a book accession number while the full records of each were kept on separate computer database or paper systems. Even these most basic functions were only available to the largest libraries because of the great expense involved. However, this was to change as computing technology became more widespread and software developed the capability to create connections between different parts of databases and to integrate functions in a primitive fashion.

As soon as it was possible to get the circulation system to 'talk' to the catalogue, it was possible to automate some library functions, such as overdue notices. As important as the mechanization of routine staff-intensive functions, was the means it gave library managers to measure and assess borrower activity in great detail for the first time through reports generated from the computer logs and databases. This knowledge greatly enhanced acquisitions and stock management strategies and has helped librarians to cope proactively with changes in funding structures, bor- rower requirements and the user community, and with the accelerating expansion in information resources.

## References

David R. Smith, (2003).*Digital Transmission Systems*, Kluwer International Publishers.

John Proakis, (2000).*Digital Communications*, 4th edition, McGraw-Hill.

Long, Paul; Wall, Tim (2009). *Media Studies: Text, Production and Context*. Pearson Education.

Sergio Benedetto, Ezio Biglieri, (2008).*Principles of Digital Transmission: With Wireless Applications*, Springer.

Simon Haykin, (1988).*Digital Communications*, John Wiley & Sons.

Timothy Binkley (1988/89). "The Computer is Not A Medium", *Philosophic Exchange*. Reprinted in *EDB & kunstfag*, Rapport Nr. 48.

# 5

# Cyberspace and the Information Society

The term "cyberspace" stands for the global network of interdependent information technology infrastructures, telecommunications networks and computer processing systems. As a social experience, individuals can interact, exchange ideas, share information, provide social support, conduct business, direct actions, create artistic media, play games, engage in political discussion, and so on, using this global network. The term has become a conventional means to describe anything associated with the Internet and the diverse Internet culture. The United States government recognizes the interconnected information technology and the interdependent network of information technology infrastructures operating across this medium as part of the US National Critical Infrastructure.

According to Chip Morningstar and F. Randall Farmer, cyberspace is defined more by the social interactions involved rather than its technical implementation. In their view, the computational medium in cyberspace is an augmentation of the communication channel between real people; the core characteristic of cyberspace is that it offers an environment that consists of many participants with the ability to affect and influence each other. They derive this concept from the observation that people seek richness, complexity, and depth within a virtual world.

The word "cyberspace" was coined by William Gibson, the Canadian science fiction writer, in his novelette "Burning Chrome" and was subsequently popularized in his novel Neuromancer. While cyberspace should not be confused with the real Internet, the term is often used simply to refer to objects and identities that exist largely within the computing network itself, so that a web site, for example, might be metaphorically said to "exist in cyberspace." According to this interpretation, events taking place on the Internet are not therefore happening in the countries where the participants or the servers are physically located, but "in cyberspace". This becomes a reasonable viewpoint once distributed services become widespread, and the physical identity and location of the participants become impossible to determine due to anonymous or pseudonymous communication. The laws of any particular nation state would therefore not apply.

The internet tools now allow remote, virtual collaboration in ways that were not possible just a few years ago. To use these tools successfully, product development teams must have a complementary IT infrastructure, matching tools, and the appropriate leadership, as well as the incentives and discipline to work together in a new way. These virtual teams - not "collocated" but "co-wired" - offer distinct advantages. Most virtual collaboration software allows for automatic recording and tracking of the proceedings of meetings. Action items are written down and indexed. Repositories of knowledge are available that make the inventory of knowledge much more visible.

Most virtual project teams have a knowledge manager position to care for this inventory. Collocated teams could take advantage of these same techniques, but few have the discipline to do so. For example, it is difficult to record and index coffeepot conversations. Partners and suppliers may be in different regions and time zones. The project can carry on through non-real time threaded discussions and overlap meetings. Shopping the world for design expertise and manufacturing capacity allows you to concentrate your design capacity on the areas where you add the greatest value. Many of these outside designers are willing to share the risks and rewards of your product, giving them high incentives for successful delivery. If managed correctly, this method of collaboration can significantly shorten time to market and time to volume manufacturing. Companies can gain a market advantage by improving their ability to do co-wired innovation faster than their competitors. The shift from a collocated to a co-wired environment requires a strategy for a Collaboration layer and a Control layer of infrastructure, the formation and training of distributed, collaborative teams, and the ability to change the way these teams are motivated and managed.

A co-wired product development team requires a communication infrastructure of two layers - a collaboration layer and a control layer. The collaboration environment includes areas for four levels of communication: project management, real-time design sharing, structured discussion and social and brainstorming spaces. The control layer builds and stores the knowledge inventory. The various design tools should smoothly connect these two layers. If they are properly organized and managed co-wired teams can develop products faster than collocated teams. Some types of communications, such as team building meetings, should be face-to-face or via video-conferencing, whereas other types such as technical exchanges can happen in real time or virtual time over the Internet, or by email. The meetings and communications should be "bandwidth-matched" to the needs at the time. Co-wired, virtual teams can take advantage of global expertise and diversity with very flexible capacity.

The infrastructure and leadership of co-wired teams must provide for fast acknowledgement of commu-nications, mutual understanding of each communication and a path for commitment and actions. The leader of a virtual, collaborative team should invest significant time in achieving the trust and social inclusion of all team members. The leader's goal is to align goals, processes, skills and tools across the team. Remote team members should be treated with the same respect and given the same promotional opportunities as collocated team members.

## Communication Resources and Information Society

There is a growing interest in using the computer to enhance instruction and learning through shared activity, and to engage students in the same intellectual and cultural activities that sustain practicing scientists and engineers in knowledge building. A major part of this learning involves engaging the student in sense-making activities, such as conversations and talk about external representations that use concepts, symbols, models and relationships.

Brown, Collins, and Duguid argue that learning involves making sense of experience, thought, or phenomenon in context. They hypothesize that our representation or understanding of a concept is not abstract and self-sufficient but rather is constructed from the social and physical contexts in which the concept is found and used. Brown et al. have emphasized the importance of implicit knowledge in developing understanding rather than acquiring formal concepts. It is therefore essential to provide students with authentic experiences with the concept. Students can engage in learning conversations in distributed multimedia environments.

The technologies allow participants to use multiple modes and representations to construct new understanding that eventually leads to conceptual change. Learning is fundamentally built up through conversations between persons or among groups, involving the creation and interpretation of communication. Learning is established and negotiated through successive turns of action and talk.

Conversations are the means by which people collaboratively construct beliefs and meanings as well as state their differences. These conversations provide a common ground or mutual knowledge about beliefs and assumptions during conversation. Therefore, one of the major issues facing designers of communication systems concerns helping one person or group understand others and create and maintain common ground.

Suchman studied the ways writing and drawing activities interact with conversation. Such activities can be used to display understanding, facilitate turn taking, and in general serve as appendages to the verbal conversation. Some experimental video systems such as Videodraw have been used to create representations that express ideas.

Some studied how design emerges from social interactions among individuals and groups as they communicate to establish, maintain, and develop a shared understanding. He observed four practices that participants used to achieve design communication:

— negotiating understanding,

— tailoring communication,

— preserving ambiguity, and

— manipulating mundane representations.

Minneman noticed that: Talk, gesture, sketching, lists and tables, formal drawings, calculations, video, photographs, and embodiments all show up as contributing to the representational and communicative activity in group designing.

These studies suggest that a multi-channel communication system provides more information and better facilitates communication between users. However, several research studies show little or no evidence of a specific benefit of multi-channel communication to outcomes of a problem solving group. Chapanis found that restricting channel capacities had little impact on the outcomes of problem-solving tasks, but did influence the process through which the result was obtained. In educational environments, this distinction can be critical, as students are expected to learn how to work in a collaborative fashion.

Various forms of computer conferencing, coordination tools, and on-line knowledge databases are currently being explored to help augment learning activities. Computer-supported cooperative work (CSCW) systems are designed to provide an interface to a shared environment in which users are linked in multiple ways such that they will perceive themselves to be communicating as if they were in the same space. Using CSCW systems, collaborators make use of a host of tools such as networked chat and draw systems, file transfer, electronic mail, and audio- and video-conferencing to work together to solve problems. When applied to educational environments, students can communicate with collaborators from institutions worldwide; the students will experience and have to contend with, cultural, ethnic, knowledge, and other differences as if they were meeting face-to-face.

Advocates of CSCW systems claim that these technologies can be used to support conversations and enhance communication. Even for people involved in face-to-face communication, images, graphics and text can be used to support the process. The use of multiple representations and multiple communication channels provide opportunities for helping one person understand the other. The studies also suggest that there is a fit between the media and a particular task or social context. Designers of CSCW systems have to be aware of the complex nature of the communication process and the use of tools and channels to facilitate communication. As designers develop new communication tools, they need to be aware of how the CSCW system fits into existing workspace. Finally, designers have to be aware of the ways people experience or perceive these new media for communication.

In an effort to help improve the teaching of concurrent collaborative design and to help students learn to work in a collaborative context the Interactive Multimedia Group (IMG) and the College of Engineering at Cornell University, have been working on a multi-year project to implement and test a networked multimedia environment.

With the advent of any new technology, there usually follows a gestation period marked by evolving features and functions. Initially, the new technology replaces rather than improves the old technology, doing the same thing but perhaps faster or on a larger scale.

Computer mediated communications (CMC), for example, have been implemented largely as expensive teleconferencing systems. Eventually, however, CMC technology will come into its own as discover how to utilize the unique attributes of this medium for supporting change and improvement. In addition to the technical design challenges of collaborative and information systems, researchers should also address the social and psychological aspects of using on-line resources and collaboration.

The ideal collaboration system would support the advantages of social, psychological and technological information resources, including communication and data access tools, on-line mentoring and experts, multiple representations for clarity, and archived exchanges and records. However we can see that students do not make effective use of the database resources and communication tools. Users need training in communication and information seeking to effectively use these systems.

In the future, CMC technology will have a profound impact on how approach and engage in the processes of education and communication. Because group interactions are often unpredictable, need to design flexible systems that can readily adapt to change. By monitoring the individual, social, and cultural forces at work as users interact with these new systems, able to develop tools, programs, and technologies that are more responsive to the changing needs of individuals and to the changing circumstances in society.

The new media are a lot like that. "What generally stand in the way of people's understanding of new media are the very terms media and medium. As commonly used, those terms are misnomers that block understanding." Simply don't confuse a medium with its Vehicles. Very few people understand the new media simply because what most people think are media are actually vehicles within a medium. Magazines aren't media nor is a magazine a medium. Television isn't a medium nor is radio nor are television stations media. A personal computer connected to the internet isn't a medium and the many computers that comprise the internet aren't media. Neither is the World Wide Web a medium nor is e-mail a medium nor is the internet itself a medium.

Newspapers, magazines, television, radio, telephones, billboards, personal computers, the internet, the World Wide Web, and e-mail all are vehicles for conveying information within a medium or media. They aren't the media or a medium in which they operate. To understand the difference between a communications vehicle and a communications medium, merely need to understand how the terms medium, media, and vehicles are correctly used when discussing transportation. Indeed, you will then also understand that only three communications media exist, what those three are, and how to use them. Only three transportation media exist.

Land was the aboriginal transportation medium; it was the first transportation medium. Humans have has walked on it since time immemorial. We still do. But we've also built vehicles to help convey us in this medium: carts, chariots, carriages, bicycles, trains, automobiles, trucks and lorries, etc.

Water is the second transportation medium. Its use as a transportation medium is almost as old as humanity's use of land, dating from whenever the first human attempted to ride a floating log or to swim across a stream, river, or lake. We've since created vehicles to convey use in this medium: rafts, canoes, barges, sailboats, ships, submarines, etc.

Before list the third transportation medium, note some characteristics of these two transportation media, because find that these characteristics have analogues in communication media:

— Note first that humans' use of those two ancient transportation media predated technology. The vehicles that human technology created have merely extended our speed and carrying capacities in those media.

— Also note that humanity's uses of these two media aren't necessarily dependent upon technology; most of us can walk and swim without any technology.

— And note that each of the vehicles for these media are limited by its medium. Trains don't operate on water nor do steamships operate on land.

Indeed, land and water have mutually exclusive transportation characteristics and reaches, mutually exclusive advantages and disadvantages. A person who needing transportation had to pick one or the other of these media based upon where that medium reached or upon that medium's carrying capacity.

For examples, water vehicles have almost global reach but not to landlocked cities. Land vehicles can deliver door-to-door, a capability that water vehicle can't provide. But many water vehicles have much greater carrying capacities than do land vehicles.

Throughout most of human history, people were limited to those two transportation media and those media's mutual advantages and disadvantages. A third transportation medium was inconceivable. But in 1783 two French brothers named Montgolfier used their era's technology to vehicle that transported them into an entirely new medium. A hundred years later, Otto Lilenthal fabricated the technology of airfoils and began gliding over the German countryside, ponds, and lakes. And 20 years after that, two American brothers named Wright determined how to marry an engine to a glider. You probably know the rest.

Utilizing technology, these pioneers opened a third transportation medium that until then had been little more than a dream — the Sky. Other vehicles developed for this new transportation medium are balloons, parachutes, gliders, airplanes, helicopters, and lately, spacecraft. These are vehicles that can transport people anywhere on Earth. Though the transportation media of land and water have mutually exclusive reaches, this new transportation medium of the sky encompasses the reaches of both land and water, and generally without the complementary advantages and disadvantages of those two prior media. But note that this new transportation medium of the sky is entirely dependent upon technology, unlike the two prior media. The sky isn't a natural medium for humans; people can walk and swim but we cannot fly.

As with transportation media, two of those communication media are ancient and arose independent of technology. But the third medium is relatively new and its use is totally dependent upon technology. Oddly, the first and earliest of these three communications media is only one not to have a commonly accepted name. So, we'll call this first medium the interpersonal medium. This aboriginal medium arose in basic animal communications, predating both humans and technology. Human technology later extended its speed and reach. Interpersonal conversation is the basic form of this medium. The vehicles that human technology later built for it include the postal letter, telephone call, and electronic mail.

Just as the transportation media of land or water have some unique characteristics, so does the this interpersonal medium of communications.

It notably has two hallmarks:

— Each participant has equal and reciprocal control of the content conveyed.
— And the content can be individualized to each participant's unique needs and interests.

However, those hallmark advantages come with equal disadvantages:

— The equal control and also the individualization of content degrade into cacophony as the number of participants increases beyond two, for example, try simultaneously holding different conversations with more than one person.

For those reasons, this interpersonal medium characteristically is used for communications between only two people. And why many academics who study communications media term it the 'one-to-one' medium. The mass medium is the second communications medium. Most people mistake the mass medium as a product of technology and don't realize how old it really is. Like the interpersonal medium, the mass medium predates technology. It originated with the utterances and speeches of tribal leaders, kings, and priests. Technology has merely extended its speed and its reach to global dimensions. Some vehicles in the mass medium are edicts, oratory, sermons, scriptures, plays, books, newspapers, billboards, magazines, cinema, radio, television, bulletin boards, and webcasting.

Communications in the mass medium generally go from a one person, for examples, a leader, a king, a priest, a publisher, or a broadcaster to many people (the audience, readership, listenership, viewership). This also is why many academics who study communications media term it the 'one-to-many' medium. The hallmark characteristics of the mass medium are:

— That the same content goes to all recipients.
— And that the one who sends it has absolute control over that content.

The corresponding disadvantages of the mass medium are:

— That its content cannot be individualized to each recipient's unique needs and interests and that the recipients have no real control over that content.

Like the interpersonal medium, the mass medium isn't necessarily dependent upon technology. For example, an actor or speaker can perform without any technology.

Before we list the third communications medium, let's note some contrasting characteristics of these two earlier communications media. Just as the transportation media of land and water have mutually exclusive characteristics, so do the interpersonal medium and the mass medium for communications:

— The interpersonal medium can deliver an individualized message but only to one person at a time.
— The mass medium can simultaneously deliver messages to an infinite number of people but its messages cannot be individualized for each recipient.
— The interpersonal medium allows each participant equal control over the content.
— The mass medium allows control over the content by only one person.

Those mutually exclusive characteristics of the interpersonal and mass media have been important because anyone who wants to individually communicate a unique message to each recipient has had to use the vehicles of interpersonal medium. And anyone who wants at once to communicate message to a mass of people has had to use the vehicles of the mass medium.

Just like using the sky as a transportation medium, for most of human history the possibility of any third communications medium existing had been inconceivable. Anyone needing to communicate had to choose between the mutually incompatible characteristics of the interpersonal and the mass media. But, Just like how several technologies converged nearly a century ago to make the sky a transportation medium, the evolution of several ostensibly unrelated technologies converged during the past century to create a third and entirely new communications medium. Among those convergent technologies were:

— The invention of digital communications during the late 1940s;
— The invention of the Transport Control/Internet Protocol ((TCP/IP) in the late 1960s;
— ARPANET's creation of the Internet during the early 1970s;
— The invention of the personal computer in the late 1970s

To lesser degrees of the importance:

— The invention of the HyperText Transport Protocol (HTTP) in the late 1980s;
— The opening of the Internet to the public in 1992;
— The invention of the Mosaic browser software in that same year.

These and other technological innovations converged to create a new communications medium that has characteristics inconceivable even a decade ago.

The hallmark characteristics of this new medium are:

— That individualized messages can simultaneously be delivered to an infinite number of people.
— And that each of the people involved shares reciprocal control over that content.

In other words, the new medium has the advantages of both the interpersonal and the mass media, but without their complementary disadvantages. No longer must anyone who wants to individually communicate a unique message to each recipient have to be restricted to communicating with only one person at a time. No longer must anyone who wants at once to communicate message to a mass of people be unable to individualized totally the content of that message for each recipient. Note that the new medium for communications, like the transportation medium of the sky, is entirely dependent upon technology, unlike the two preceding communications media. Like humans flying with technology, this form of communications can't be done with technology.

Because the new medium simultaneously encompasses both the characteristics and the reach of the two previous communication media and therefore can easily perform each of those media's individual tasks, many people mistake the new medium as merely an electronic extension of the interpersonal or mass media. Most people mistake it as a paperless or antenna-less form of the mass medium. Moreover, many marketing consultants often mistakenly refer to it as a 'one-to-one' medium.

But the academic and the consultants who truly understand this new medium and its possibilities to simultaneously deliver an infinite number of individualized messages while providing equal control over that content refer to the new medium as the 'many-to-many' medium — to distinguish it from the 'one-to-one' (interpersonal) or 'one-to-many' (mass) media. Mistakes, misnomers, and misperceptions of the new medium are easy to make because the vehicles of this new medium are only starting to appear, as are the true capabilities of this new medium.

Just consider the converged technologies that make this new medium possible. For instance, the millions of computers interconnected through the internet can acquire, sort, package, and transmit information in as many ways as there are individual people. They can establish those communications simultaneously. And they allow each participant to share equal simultaneous control. This can result in unprecedented forms of communications.

Imagine that when a person visits a newspaper Web site, he sees not just the bulletins and major stories that he wouldn't have known to request information about but sees the rest of that edition customized to his own unique needs and interests. Rather than every reader seeing the same edition, each reader sees an edition that has simultaneously been individualized to his interest and generalized to his needs. Or imagine that each viewer who is simultaneously watching a broadcast can stop, rewind, or fast forward the program at will, or even change the denouement of the program's plot. Realize that these new medium forms of content inherently are forms of mass customization, something impossible with either the interpersonal medium or the mass medium. The existence of this new medium will catalyze, economize, and popularize entirely new vehicles for production and distribution, just as the invention of the medium of air did for transportation.

The new medium itself is merely a manifestation of a larger, revolutionary historical change underway that transcends just issues of communications. Analysts and pundits talk about this larger change as an Informational Revolution that is superceding the world of the Industrial Revolution in much in the same way that the Industrial Revolution itself superceded the Agrarian Revolution. But the terms Informational, Industrial, and Agrarian merely refer to characteristic products of these phases in the evolution of human society. Not to the natures of these revolutions themselves.

During the Agrarian world, which began about 5,000 to 10,000 years ago, a person who needed clothing had to make his clothing himself. Or another person made his clothing individually for him. Only one piece of clothing would be made at a time and each piece of clothing would be individualized to his own individual size and needs.

Then some 200 years ago, mechanical technology evolved to the point where factories could created, and the Industrial Revolution began. In the Industrial world, millions of pieces of clothing could be produced, but none could be individualized to the exact measurements of its ultimate consumer. Note how these complementary advantages and disadvantages of products from the Agrarian and Industrial ages are similar to the complementary advantages and disadvantages of the interpersonal medium and the mass medium.

This is because most of what we nowadays perceive to be the interpersonal medium dates from the Agrarian Revolution and most of what we nowadays perceive to be the mass medium dates from the Industrial Revolution. Now, new advancements have created technologies that unite the advantages of both Agrarian individualized production and Industrial mass production, with none of the complementary disadvantages. For example, companies using these new technologies,such as Levis Strauss & Co., have begun to mass-produce jeans that are individualized to each users exact specifications. This is similar to how the new medium can send instantly individualized messages to mass of recipients. Both the new medium and Levis Strauss & Co.'s abilities to mass produce of individualized jeans are manifestations of the Informational Revolution.

The Informational Revolution's effects upon society are being compared to those from the invention and promulgation of printing presses. However, the actual significance of the Informational Revolution is greater: The invention of the printing press was merely a technological amplification of the mass medium. By contrast, this new medium indeed is an entirely new medium; a quantum leap beyond mere innovations such as the printing press.

Returning to our analogy between transportation and communications media, the development of the air as transportation medium didn't entirely replace land or sea transportation. Neither will the development of this new communication medium entirely replace the interpersonal or the mass media. However, it will certainly and markedly reduce and limit those previous media, much as the invention of aviation did to land and sea transportation.

## Communicative Interactions

The public/private distinction is a complex one, which in modern capitalism is so often confused by the extension of private control into the 'public sphere' as market place. The traditional pre-capitalist market place is not a place of private interests negotiating, but of the public good of exchange. Today the private exists in the public sphere as can the 'public' exist in the private. Privacy might be commonly thought of as being confined to the spaces of the home, but this is also increasingly the place where, paradoxically, individuals gain access to the public sphere. The fact of this is mutually generated: the less individuals engage in practices of interaction in 'public spaces', the more they are likely to be engaged in interactive practices in private spaces, and vice versa. Under these conditions, the household unit becomes a primary cell of modern social relations, the basic unit and building block from which social interaction occurs. When the public sphere has withdrawn to the home, where a 'dialogic' or two-way open interaction becomes impossible, interaction becomes more and more 'confined' to the family, the household, and where one works.

These conditions certainly did not obtain in pre-media society in which the frequency and intensity of embodied interaction are of an entirely different order. The origins of European modernity since the 18th century, are founded on the café as the bedrock of the emergence of a public sphere.

However, in media societies where the geographic and kinship ties of the parish, local neighbourhood, or the industrial slum have virtually disappeared, individuals have historically become very heavily dependent on media of many kinds to acquire a sense of belonging and attachment to others. The situation is one of separation and unity. Individuals are separated at a geographic level, locked away in their house allotment or unit fortresses, but united on scales of city or nation in their attachment to forms of media.

Ironically, the marketing calls for consumers to 'get connected' and 'travel on the Internet' instead of being 'stuck at home' are an exact reproduction of the social and urban consequences of broadcast technologies. On the one hand, individuals are told they can interact to overcome the tyranny and restraints of broadcast, but they do so only by reinforcing the domestic conditions of their atomised existence.

The question of whether interaction, once it is reduced to the electronically mediated and technologically-extended kinds of access to communication which is enabled from the home, constitutes participation in a public sphere is a pivotal one to ask in relation to CMC. Certainly the private/public question becomes extremely vexed on the Internet. Some suggests that "If 'public' discourse exists as pixels on screens generated at remote locations by individuals one has never met and probably will never meet, as it is in the case of the Internet with its 'virtual communities', 'electronic cafes', bulletin boards, e-mail, computer conferencing and even video conferencing, then how is it to be distinguished from 'private' letters, printface and so forth."

Symbolically, as well as functionally, the cybercafé is extremely interesting. It strongly reaffirms the idea that the cellular network basis of gaining access to the public sphere predominates, where even one of the strongest institutions of embodied public life can be remade in terms of CMC. Nobody meets face-to-face at a cybercafé, as the face-to-screen precludes dialogic contact in any form other than electronic.

To understand whether electronically-mediated communication, of either the broadcast variety or network form, is capable of constituting a public sphere, the difference between these forms needs to be discussed. The contention that the Internet, and CMC in general, have radically altered the nature of public communication is largely built on the assumption that they are progressively coming to usurp the power of broadcast media.

Theorists of the second media age argue that both broadcast and interactive communication apparatuses have together constituted the primary forms of cultural mediation in information societies since the second world war. The important point here is that it is not possible, in this view, to understand the second media age without first understanding the first media age. The dependence on media of both first and second media are interrelated. However theorists of cybersociety such as George Gilder in Life After Television, Sherry Turkle in Life on the Screen, Mark Poster in The Second Media Age, contend that the second media age has arisen on the back of the conditions produced by the first. These conditions, the production of an indeterminate mass by broadcast, the separation of individuals from the means of producing their own contributions to public communication, and the disintegration of traditional community, are all hailed to be overcome by the Internet.

The Internet is, above all, a decentralised communication system. Like the telephone network, anyone hooked up to the Internet may initiate a call, send a message that he or she has composed, and may do so in the manner of the broadcast system, that is to say, may send a message to many receivers, and do this either in 'real time' or as stored data or both. The Internet is also decentralised at a basic level of organisation since, as a network of networks, new networks may be added so long as they conform to certain communications protocols.

According to the second media age perspective, the tyranny that is attributed to broadcast lies in its hegemonic role in the determination of culture as well as individual consciousnesses, which derives from its predominantly vertical structure. This structure is one in which the individual is forced to look to the image and the monopolisation of information and entertainment to acquire a sense of assembly and common culture.

The second media age, on the other hand, bypasses this 'institutional' kind of communication and facilitates instantaneous, less-mediated and two-way forms of communication. For the romantic variety of cyber-utopians, on the other hand, it 'restores' such communication.

At the level of interaction, the second media age utopians point out the empirical increase in the take-up of the Internet and other network technologies, and the fact that empirically it is true that the Internet is mainly interaction and very little broadcast whilst television is mainly broadcast with very little interaction, as evidence for the 'ontological' nature of the second media age as a distinctive trend, movement and modality of social integration.

The importance of the fact that the many can interact with the many in cyberspace is almost exclusively related to the way it is said to break the 'lock-out' predicament which individuals face in broadcast interaction. The walls that are erected by the power of broadcast rapidly disintegrate as a form of electronic communication is made available which is adequate in speed, form and complexity to the abstractness of the social forms which have denied mediated interactivity by way of broadcast.

In broadcast communication, the individual receiver of messages is subject to one-way communication from the 'elite' producers of messages. The horizontal connection with other consumers of the same messages, is generally only possible via the fetish of the image or the celebrity, in whom 'concrete consciousnesses are concentrated'. Conversely, with the Internet, the message producers are by-passed, as the walls that are erected at the horizontal level effectively disappear.

These 'media' walls are the result of the architecture of broadcast itself. The more the individual looks to the media for acquiring a cultural identity the less they look 'sideways' for interaction. Conversely, the less the individual looks sideways for social solidarity and reciprocity, the more this mode of association becomes weak and de-normalised, and so the alternative dependence on a centralised apparatus of cultural production becomes imperative.

In the second media age, however, the walls separating individuals at a horizontal level are said to be overcome, as the individual looks directly to others for a sense of millieu and association.

The Internet lifts individuals out of the isolation created by media walls - particularly as these walls are reinforced in urban contexts. In information societies, individuals increasingly interact with computer screens, developing face-to-screen relations rather than face-to-face relations, but this opposition is no longer significant, some argues, when the larger cultural contexts of post-industrial societies are eroding the boundaries between the real and the virtual.

The extent to which the Internet is hailed as an overcoming of fragmentation and individualism is quite remarkable in recent literature. In some cases it is attributed with an integrative function which is able to correct a tendency that is over two hundred years old. The message of redemption which is promoted in the second media age thesis, be this for public or private, is a resounding one - a message whose dreams of unity has theological undertones.

The success of any argument claiming a special role for the Internet in the constitution of a new public sphere rests on its ability to establish an imaginary unity in which all participants have equal opportunity for 'observation' and communication. This postulated imaginary unity, most well known in the phrase 'virtual community', seldom reconciles itself with the fact that 'the Internet' is not at all technically homogenous and is segmented into quite a range of properties and capabilities, each of which carry different sociological and communicative potentials and effects.

It is true that, unlike television, the Internet is a network as well as 'dialogical' capable of a two-way dialogue, and for this reason is applauded as a universalising structure of communication. But its network properties are rarely realised in communication directly, rarely do they become meaningful qua network because, individuals only ever 'use' the Internet within well-defined sub-mediums. Trevor Barr usefully breaks down the different kinds of interaction on the Internet into six categories:

— one-to-one messaging (such as email);
— one-to-many messaging (such as 'listserv');
— distributed message databases (such as USENET news groups);
— real-time communication (such as 'Internet Relay Chat');
— real-time remote computer utilisation (such as 'telnet'); and
— remote information retrieval (such as 'ftp', 'gopher' and the World Wide Web').

It can be seen from this list that the Internet provides a generic environment for a number of different modes of interaction which can vary according to real-time/stored time, symmetrical versus asymmetrical dialogue, broadcast sending and receiving and information posting and retrieval.

But each of these modes of interaction relate very differently to the possible constitution of an 'electronic public sphere'. Moreover, the information and communication possibilities of the Internet are more often than not parasitic of broadcast-mediated communication. The growth of companion websites which accompany media organisations, newspapers, consumer products, sporting events etc, provide an astonishing array of reasons why information retrieval, listserv, and interactive databases available on the Internet are turned to.

When CMC is broken down into specific sub-media's rather than reduced to the indeterminacy of 'the Internet' as a communication environment, a more sophisticated appreciation of the change technological transformations of the public sphere is enabled, and the advancement of new accounts of context-specific partial publics is one outcome. However, at the same time, the global reach and mobility of all forms of Internet communication, regardless of the specificities of their sub-media also needs to be accounted for.

Why this is significant is that, whereas broadcast generates an instant 'international context' of social connection, there are few ways in which individuals can achieve meaningful interaction to make tangible these global connections. There are telephones and other 'narrow-band' ways of communicating, but none of these is quite able to provide a multi-media context for any given interaction. The Internet, it is argued by its promoters, changes all of that.

In accounting for the growth of CMC via the Internet, both national and global statistics become significant. Given that the experience of community on the Internet is not limited to national boundaries, the shape and structure of this virtual community becomes significant. In 1996, five years after the Internet became fully commercially available, only 7% of Australian homes had connected to the Internet, whilst 25% already had access in their workplace.

In 1998, the domestic connection figure had almost tripled to 19%. The number of Internet users worldwide at the end of 1998 was 147,800,000 of which 52% were American. World wide, therefore, the number of Internet users had increased from approximately 25 million in July 1995 - almost a sixfold increase.

Besides being hailed as a technology which can deliver the 'global village'. The Internet is also promoted as a singular medium which allows for democratised processes which were not previously possible in the era of broadcast. But what kinds of democracy are being postulated here? Traditionally, and more than ever now, democracy is heavily aligned with the nation state. Because of this, nonsense is made of the claim that the Internet enables universal participation in democratic process.

The point here is that practices of communication afforded by CMC may be able to substitute some of the functions of the mass media - for example in the formation of pre-instutional public opinion' - but do not necessarily exert pressure on the institutional apparatuses of politics. Of course the mass media itself, as a means of electronically mediated communication, can never replace the institutional apparatuses of politics and, as numerous studies have shown, has been just as much used by politicians as it has influenced them.

The Internet can properly be classified as a 'global' technology, which enables connections with individuals and institutions overseas just as easily as is does nationally, regionally or locally. If there is an imagined community on the Internet, it is definitely not the nation state. State-bounded kinds of citizenship cannot be considered coterminous with the kinds of citizenship which are achieved on the Internet. However, this is not to argue that a global sense of citizenship, even if it too is an 'imagined one', cannot exist. Recent protests against international financial insitutions such as the World Bank were organised almost entirely through Internet media—a case of not so visible electronic assemblies producing very visible embodied assemblies.

John. B Thompson's book "The Media and Modernity", he argues that it is not possible to arrive a satisfactory understanding of the nature of public life in the modern world with a conception of 'publicness' which is spatial and dialogical. By seeing the public sphere in only these terms are invariably obliged to 'interpret the ever growing role of mediated communication as an historical fall from grace'. Thompson argues that this is so because the widespread conception of publicness used by political and communication theorists alike is one based on mutual face-to-face relations. To problematise this, Thompson takes issue with Habermas' model of the public sphere which is criticised for being metaphysically flawed.

## Computer-Mediated Colonization

Frits Grotenhuis analyzes how cultural differences between the Dutch and Americans issue in mis-communication in Computer-mediated Colonization (CMC) environments. This not only frustrates the goals of business mergers, it further documents important ways in which CMC technologies of themselves cannot overcome cultural differences. Dineh Davis turns attention to a much broader sense of culture—the culture of the Internet and its technologies as favoring the extraverted, the busy, and the noisy, in contrast with the personal and cultural preferences of much of the world for ways of being and becoming human that stress humility, quiet, and reflection.

Grotenhuis and Davis thus complement one another as Grotenhuis demonstrates cultural conflicts external to CMC technologies, while Davis documents cultural conflicts embedded in these technologies and their implementations in a global Internet culture. Finally, Alexander Voiskounsky's research suggests an interesting and creative middle ground, i.e., multiple uses of the Net and the Web that utilize Russian as a lingua franca and thus helping people from diverse countries sustain and foster important cultural activities.

As Grotenuis makes clear, culture plays an ongoing role in the "nuts-and-bolts" elements of international business—first of all, in efforts to merge companies from diverse cultural settings. In particular, contra the assumption that CMC technologies will somehow ease these processes, culture rather complicates the use and efficacy of CMC technologies. This brings to the foreground a crucial lesson: some problems such as "communication failures" or "lack of teamwork" are better understood as breakdowns in intercultural communication. At the same time, of course, success for managers in a global society depends on their ability "to communicate successfully with managers and employees from other cultures."

Grotenhuis explores the impact of culture on two Dutch-American mergers by way of focusing on "cultural collisions" related to CMC practices, using a case study approach. Some fascinating differences emerge here between the two cultures in terms of business. Americans tend to be short-term oriented and result-driven. Americans are in a "permanent process of moving, of instability", which is perhaps not surprising in a country with few unions, in contrast with the trade unions and working councils in the Netherlands.

Grotenhuis further notes that the brevity of e-mails, while favored by the medium and "netiquette", contributes to culturally-based miscommunication. Americans tend not to provide context, which makes Dutch recipients feel ill-informed. Contrary to cherished American emphases on equality, informality, and direct communication, the Dutch regard their US colleagues as more hierarchical and less honest. As well, the Americans are perceived as more ethnocentric: "Generally they think that what works in the US, that will be good for Europe as well, instead of adapting to a different situation. What is good for the USA, that is not always necessarily good for Europe too." Finally, as these comments suggest, stereotyping occurred that hindered communication.

In particular, Grotenhuis notes that "despite all advantages of computer-mediated communication, still, face-to-face meetings remain necessary to prevent clashes in the long run." In this way, Grotenhuis' study stands as one more piece of evidence in the larger turn towards embodiment as an essential feature of human identity, one that is not easily abandoned for whatever advantages and freedoms may enjoy in cyberspace.

On the contrary, the risk is precisely that, in the face of the diversity of aptitudes, attitudes and values that make of "happiness" or "optimal experience" for individuals and cultures, when begin to base an entire culture on a single technological platform, are bound to create constraints on some subset of the population. When such a platform is installed cross-culturally and globally, dissent and dissonance become inevitable."

This dissent and dissonance is apparent first of all on the level of human identity, i.e., in the conflict between the tendency of the media to privilege publicity, self-promotion, busy-ness, indeed "noise," and those cultures and individuals that instead endorse humility, privacy, patience, and deliberation. As well, the role of English as a lingua franca means that the Web and the Net are limited by the limits of this particular language. Like all languages, English is good for specific kinds of expressions and thought, and less capable of expressing, or even allowing, others.

Davis further characterizes cyberspace as the equivalent of the "tyranny of the majority" over a voiceless minority - "... those whose biological, cultural, or behavioral upbringing precludes them from embracing the exhibitionist, impatient mentality of the other half of the human race," as this latter is at work in the technoculture of cyberspace. Indeed, the very term "cosmopolitan" - contra its claim to universality - excludes a significant portion of humanity:

The very concept of cosmopolitanism can have tremendous appeal to some, while it may not serve the emotional needs of others who have led happy, albeit sheltered and "provincial" lives, in isolated communities. Though a culture of consumerism may appeal to some, it has brought others closer to an understanding of how obvious disparities can become across the producer-consumer divide. If colonization and invasion are inevitable then, she hopes, the colonizers will learn the lessons their predecessors - i.e., that conquest is more secure and stable when elements of the old culture are preserved in the new culture, if in albeit transformed guise.

Davis' initial summary of the presumed values and orientations of technoculture include: "all humans have the same intrinsic interests and value system based on a Protestant work ethic, a need for higher challenges and matching skills, a desire to move from physical to mental labor, an affinity for abstraction and compartmentalization of life affairs, and a peak pleasure experience in an intellectual, disembodied state." Again, this critique of disembodiment in cyberspace vis-à-vis embodiment and "non-mediated interpersonal relationships" contributes to the central debate and recent turn in CMC literature.

Contra utopian hopes that cyberspace would prove to be "gender-blind" and thus a space in which women could enjoy greater equality and acceptance, an extensive body of research has now made it clear that carry gender with us into cyberspace - sometimes with terrifying results, including the famous "virtual rape" in cyberspaceand other forms of discrimination and violence towards women.

Gender, of course, is a central element of the worldview defining specific cultures, i.e., the roles played by males and females, their relationship to one another (hierarchical, egalitarian), etc. It is hence perhaps not surprising, except for those holding to the more utopian view, that gender differences pervade attitudes towards and uses of CMC technologies, even in cross-cultural comparisons. Nai Li and Gill Kirkup's study of British and Chinese students' use of and attitudes towards the Internet. Despite the considerable differences between British and Chinese cultures, they share similar patterns of gender differences, ones that work to the disadvantage of women.

Ann Willis makes clear that these patterns of gender discrimination are at the heart of North American Internet culture as well - at least as that culture is represented by Wired magazine, the putative voice of the "digital revolution." Here again, despite ostensible commitment to radical equality as a primary goal of the digital revolution - indeed, a goal and value that function as crucial arguments justifying or legitimating the revolution - Wired instead works to "re-power" the familiar elites of earlier epochs, i.e., white males in the middle and upper classes.

Li and Kirkup begin with a helpful review of the pertinent research on gender vis-à-vis information technology. The findings are not encouraging for those committed to gender equality: women have more negative attitudes regarding computers and greater computer anxiety; women use these technologies less, and are less likely to take computer courses or pursue computer-based careers. These differences extend to attitudes towards and use of the Internet in numerous ways, and this, despite the fact that women now form the majority of Internet users. As Li and Kirkup point out, much of the research on gender and the Internet has been done in Western countries, but it is also clear that attitudes towards computers and CMC technologies depend in some measure on culture.

Willis notes that the hype surrounding the 1990s "digital revolution" - more carefully, the utopian, specifically libertarian vision of individual liberation in cyberspace as free market— is dying down, in part as it is overshadowed by more familiar stories of "old and new media mergers, deals and machinations as companies re-shuffle to position themselves within an on-line services communication context."

In this context, perennial questions remain regarding access, identity, power, etc., in sharp contrast with the utopian promises of radical equality and democracy. In Willis' view, this is not surprising as these issues were raised by what might think of as the flagship journal of the digital revolution, Wired. Simply, while Wired professed to be the "voice of the digital revolution"- one as ostensibly radical as the industrial revolution - Willis' analysis makes clear that this version of the revolution was always only for the elite, i.e., white males, college educated, at the top of the income scales and prominent institutions in the United States. This argument is extended later on in her matter as she observes that in the period between 1993 to 1996, only one African-American was pictured on the cover — and that, to illustrate not individual success in an ideal free cyber-market, but to image the hacker who inhabits the margins of a predominantly white world of success and power.

In particular, Wired emerges as the bully pulpit for laissez-faire capitalism and privatization - what Willis calls "techno-libertarian" ideology. In dramatic tension with its democratic claims, the world of Wired is first of all a gated one. By focusing on ostensibly technological solutions to political issues — e.g., increasing bandwidth as eliminating information scarcity and inequalities — this ideology sidesteps and suspends genuine issues of power, as participation in Wired's vision of the future clearly depends on social and financial status, knowledge, and education.

Indeed, as its laissez-faire credentials suggest, the magazine - consciously or not - invokes an equally 19th century Social Darwinism: "This is the digital revolution, you are either part of the steam roller or part of the road." This Social Darwinism at work in the competition of the marketplace, Willis later points out, hence provides techno-libertarians with a handy response to concerns about the digital divide, the gap between information haves and have-nots. In a Social Darwinist world, winners and losers are a given, a necessary outcome of competition, not a problem of justice or reason for concern.

This intrinsically hierarchical worldview is further consistent with Willis' observations of the marginalization of women and people of color in the world of Wired. This marginalization, Willis notes, squares with the more general observations of how such publications replicate sexist discourse despite their overt claims regarding egalitarianism as an inevitable and desirable outcome of the digital revolution. As Willis sees it, Wired instead "re-powers the binary 1950s gender ideologies of women positioned in the domestic private sphere with men in the public sphere of work and business."

This occurs in numerous ways, perhaps most prominently through the metaphor of "frontier" for online culture - i.e., "a wild, dangerous, out of control place," one where only tough guys survive and women appear, if at all, in the role of whores or good girls.

Perhaps the greatest disjunction between the revolutionary rhetoric of liberation through technology and the realities of online praxis is in the central legitimation claim for the digital revolution - namely, that it will inevitably democratize. The libertarian version of democracy, as Wired makes clear, however, is rather the (limited) plebiscite notion of "clicking For or Against" - i.e., responding to ostensible choices, absent dialogue, debate, and the hard slow work of building consensus among diverse perspectives. This latter view, however, makes sense in the ideological framework of Wired because it coheres and resonates with the free market of commerce as the primary model of human interaction. Behaviors and choices as producers and consumers, as negotiators in a marketplace, are thereby conflated with "democracy". Consumer trading is equated with public participation - one participates in "democracy" by behaving as a consumer.

This commodification of democracy, one that morphs it into a primary free market model, is part of a larger pattern of commodification that Willis calls "über-consumption." Willis cites here Solnit, who finds PacMan to be the symbolic metaphor of such über-consumption - its sole purpose, after all, as a disembodied head-mouth "was to devour what is in its path as it proceeds through an invisible maze". In the face of this analysis, Willis then asks the sensible question, "...what exactly is revolutionary about Wired's digital projection?" — especially when it appears that Wired rather works to reinscribe "older, feudalistic and laissez-fair, ideals and practices, thus inhibiting possibilities of the participatory potential for modes of communication like CMC."

In broader terms, contra the postmodern claims of the 1980s and 1990s that new means of communication automatically mean revolutionary changes in social, political, and cultural practices. Willis makes clear, rather, that culture is conservative —it is much rather the case that older patterns of belief, values, and practices will express themselves in new forms made possible by new technologies, instead of the case that new technologies will reshape those older patterns.

Willis' discussion of Wired, much has been made of "electronic democracy" as one of the specific ways in which CMC technology would fulfill utopian hopes for liberation via technology - whether in modernist or postmodernist guise. In particular, Jürgen Habermas's theory of communicative reason and correlative concept of the public sphere has provided a central theoretical foundation for explorations of how far CMC technologies might indeed serve democratic hopes. While Habermas is critical of especially libertarian views of democracy, libertarianism predominated Net ideologies in the 1990s. Tim Jordan's essay provides an extensive justification of such Net libertarianism as part of his analysis of "cyberpower."

Unlike earlier enthusiasms for life online that emphasized radical discontinuities between embodied and disembodied identity, Jordan's treatment of identity and renovated hierarchies recognizes a number of important connections between online and offline experience. So, for example, while acknowledging that certain aspects of offline identity - e.g., gender, race, geographical location, etc. - may be hidden online, Jordan claims that it would be a mistake "to assume that the powers that flow around offline identity...are absent online...." Instead, he argues, identify the particular forms of identity that emerge in cyberspace and what kinds of power attaches to them.

On this basis, moreover, he attends to the ways in which "offline hierarchies are reinvented online", precisely on the basis of online identities as still implicating offline identity. This means, in different terms, that in contrast with the either/or's characteristic of 1980s and 1990s emphases on the radical differences between online and offline, Jordan's account of cyberspace is something of a middle ground. This middle ground recognizes the both/and - the ways in which cyberspace is both different from and similar to ordinary lives as embodied human beings in real spaces. So, for example, Jordan reiterates some of the familiar claims regarding the difference between online and offline - e.g., that hierarchies will be flattened by more egalitarian forms of communication that emerge in cyberspace, and that the Net, in the famous phrase, interprets censorship as damage and simply routes messages along different pathways, thus undermining governmental efforts at control.

At the same time, however, identities in the real world, and the relative power or lack thereof that we enjoy, accompany us into cyberspace. By the same token, Jordan points out this same interrelationship between the individual and the technologies of CMC: "individual powers in cyberspace will be defined by the technology we are using and the capabilities this technology offers." This is so, moreover, because the technology embeds social and/or ethical values - an aspect of what Jordan calls "technopower," the conjunction of seemingly dead technology with living social values that "constitute the very possibility that cyberspace exists in the first place."

This technopower, more broadly, has a trajectory. Contra the liberationist theme characteristic of 1980s and 1990s postmodern enthusiasm for cyberspace, Jordan claims that "The direction of technopower in cyberspace is toward greater elaboration of technological tools to more people who have less ability to understand the nature of those tools." Instead of the technology delivering greater power and control to the users —the promise of all modern technologies—control is rather delivered "to those with expertise in the increasingly complex software and hardware needed to constitute the tools that allow individual users to create lives and societies."

This process involves information overload, as moving information from the costly and time-consuming production processes of print into cheap and quick electronic forms contributes to the explosion of information. The solution to this, of course, is more technology to organize, filter, and dispose of information. However, this only adds to the problem of information overload, according to Jordan, leaving only an elite of real experts to enjoy any putative freedom in cyberspace. Contra a "hopeful form of power" for the individual, "...cyberpower of the social is pessimistic because it reveals networks of interactions that increase the ability to act of an expertise-based elite."

Jordan's third layer of cyberpower is the imaginary, the layer that includes sense of urgency to either jump on the bandwagon of new possibilities just around the corner and/or to flee the imagined disasters of dystopian futures. Jordan nicely summarizes the various forms of utopias imagined for cyberspace, including hopes for personal immortality, virtual "heavens" in which perfect equality is achieved, and the cyborg as overcoming binary oppositions that found oppression. Favorite dystopias, by contrast, turn on cyberspace qua panopticon, i.e., the space in which the individual becomes perfectly transparent to complete surveillance. Jordan makes the important point that both these hopes and fears rest on the belief that "everything can now be manipulated through information codes," i.e., there is no resistance, nothing that cannot be reduced to information.

For Jordan, finally, the future is up for grabs. While he argues that libertarianism - the ideological conjunction of individual liberty and free markets - is best suited to cyberspace, he sees it as contested ground, a battleground between the individual and the social, including the possibility of dominance by elites.

Holmes begins with a review of some of the most important voices regarding democratization in cyberspace, including Mark Poster's seminal analysis of the Internet as facilitating especially postmodernist conceptions of democracy, before turning to Habermas's conception of the public sphere proper. He points out that the public/private distinction presumed in this conception becomes problematic, first of all, as broadcast media override geographic and kinship foundations of one's sense of the public sphere.

As well, as the Internet and other electronic media introduce interactivity - a point stressed by Poster as marking out "the second media age" - the question arises as to whether broadcast and network forms of communication make a new kind of public sphere possible. In contrast with the enthusiasts who find network communication thus redemptive, Holmes examines what kinds of interactivity are indeed made possible with CMC technologies. While he finds important mismatches between the Habermasian requirements for the public sphere and the kinds of dialogical reciprocity that may take place in CMC, Holmes likewise finds that the postmodern alternatives are also lacking - first of all, because they fall into an overly simple technological determinism.

Holmes comes to a complex middle ground, claiming first that as individuals engage in a variety of communicative media, they participate not so much in a "pre-given public sphere" but rather "in the process of constructing publicness across a range of mediums." Hence it is less the case that, as postmodernists tend to argue, "...the contemporary public sphere is breaking down and becoming fragmented as is the fact that it is sustained across increasingly more complex, dynamic and global kinds of communication environments."

In sharp contrast with the euphoric enthusiasm for the digital revolution of the 1990s, contributors thus raise a range of critical concerns. CMC technologies do not serve as a culturally transparent medium that somehow eliminates cultural differences and culturally-based obstacles to communication On the contrary, CMC technologies embed specific communicative preferences that work against effective cross-cultural communication. Indeed, CMC technologies may embed preferences for the public, the busy, the noisy, the unreflective and the self-promoting that runs directly at odds with individual and cultural preferences for privacy, silence, reflection, and humility.

In fact, both cross-culturally and at the heart of the North American self-promotion of CMC technologies, find a variety of clear biases against women. These biases, indeed, are consistent with more libertarian understandings of the sorts of freedom and democracy facilitated by the Net and the Web. Even such an articulate defender of Net libertarianism as Tim Jordan recognizes that in the competition for multiple forms of power available in cyberspace, traditional elites and hierarchies may win out once again, despite justifications to the contrary that claim greater equality and democracy will inevitably emerged in a Wired world. Indeed, the problem of realizing any form of electronic democracy is complex. The strong theoretical foundations provided by Habermas are fundamentally challenged by Luhmann, who, as taken up by Moeller, argues instead that participation in cyberspace is the end of "individuality," much less "democracy."

It would thus seem safe to say that the bloom is indeed off the revolution. But this is by no means to argue that the revolution has failed. On the contrary, that it is precisely through recognizing these critical limitations and dangers in CMC technologies that are thus able to make use of them in informed ways that are more likely to succeed in their goals. In other words, precisely by recognizing the role of culture in communication above and beyond what CMC technologies can - and cannot - accomplish, are better able to teach and train people to use these technologies in culturally-informed ways that are more likely to facilitate rather than frustrate cross-cultural communication. Similarly, by recognizing how far CMC technologies may favor male ways of knowing and communicating, women and men who are interested in more egalitarian relationships and communication can use these technologies in ways more likely to foster rather than hinder such equality. Finally, by refining theoretical understanding of what democracy might mean, and determining - through both theoretical debate and reflection on what happens in praxis as diverse users take up CMC technologies in diverse cultural settings - how far these technologies might indeed encourage more democratic forms of polity are better prepared to make effective use of these technologies for the sake of genuine democracy rather than its counterfeits.

In all these moments, in fact, it appears that if wish to realize at least some of the goals of the "digital revolution" - greater gender equality, greater democracy, etc. return to a broader understanding of how lives online rest on and interact with lives offline, i.e., as embodied human beings inextricably implicated in multiple webs of relationships with other human beings in diverse communities, traditions, histories, cultures, and, indeed, with a larger ecological system. Such a (re)turn, will include attention to the social context of use of CMC technologies as a way of shaping their design and use in light of the values and bias they can embed, and in order to avoid various forms of computer-mediated colonization, i.e., the imposition of unwanted and alien cultural values and communicative preferences that otherwise is likely to result from the naïve assumption that these technologies are culturally neutral. While there are, unfortunately, a range of examples of such colonization—there are also examples of utilizing CMC technologies in ways that avoid colonization, however inadvertent, while sustaining both local cultural values and preferences alongside communication with those beyond the boundaries of one's immediate community and culture.

## Attractiveness of Cyberspace

The Internet is obviously international and transnational. Since it first came into life in the USA, the majority of the Internet "aborigines" - beginning with the times of ARPANet - reside in North America. Moreover, the great majority of the WWW content information is located on the US/Canadian web-servers; userful data are often mirrored elsewhere. The cyberculture - whatever that means - originated in the USA. The majority of surfers and site-visitors, e-consumers, Listserv and Usenet subscribers, online chatters and players, fans of club-like web activities, discussants at webforums, etc. used to also belong to the North American culture. The newly emerged global influential Cyberculture used to be strongly dependent on a good command of English language, and on the willingness to share cultural norms established by native English speakers.

Parallel to a global view on evolving virtual communities, a culture-specific approach seems to be no less needed. "The Internet and local cultures both determine each other". The dialectic idea of interdependence between a global cyberculture and local network subcultures is principal and worth discussing. Since the global influence is quite obvious, it is time to stress the impact of local cultures on the evolving world-wide Internet culture. At first, might restrict ourselves to purely ethnic subcultures - Far Eastern, Latin American, Eastern/Central European, Southern/Central/ Western/Northern African, Southern Asian, etc. At the moment, some of them can hardly be called influential, but times change.

The advance of the new millennium marked the time period when the proportion of North Americans and native English speakers on the Net decreased to only less than the half of the global Internet population. By the end of 2000 "the US share of global Internet users dropped from 40 percent to 36 percent ... while 12 percent live in the world's other English-speaking countries. Thus, the total proportion of native English speakers decreased to less than 50 per cent: even if add New Zealand, the same would be true now in 2002.

This news is sort of a challenge for the Internet aborigines. The same source stresses that "in every global region where English in not the main language spoken, nine-in-ten Internet users prefer to get local information in their local language". Of no less importance are customers' orientations: "Web users are up to four times more likely to purchase from a site that communicates in the customer's language. Are the North American high-tech producing, consulting, and/or trading companies going to save their once-privileged positions in the rapidly changing market? If local markets are important, the companies' CEOs should pay special attention to multinational communication, to advertisements, to precise and culturally adequate translation of news, products and services presented on websites.

Thus, the Internet is now multilingual and multifocused. Taking into consideration non-English-speaking language-specific ethnic groups, one should mention that some of them have developed web sources that look attractive for both residents and foreigners. The volume of content based on the use of the Spanish language seems to be quite sufficient not to bother with English-based web content. Even for bilinguals familiar with the English language, non-English web-based chats and information sources oppose or substitute the global Internet.

Some countries pursue protective policies towards the integrity of national culture and ethnic language in Cyberspace, for example France or

China. Sometimes ethnic segments of the Internet originated from abroad. The initial Arabian newsgroups and listservs, for example, were organized by emigrants from different Arabian states. More often representatives of diaspora communities participated in web activities initiated in metropolias. This is highly characteristic for the USA. The Global Internet Statistics are estimated that "there are some 16 million Americans who are more likely to access the Internet at home in their own language, rather than in English", while in the office they are more likely to access the Internet in English. The leading languages opposing English in the USA as a medium of home use of the Internet are, in descending order, Spanish, Chinese, Italian, German, Korean, Polish, Japanese, Portugal, and Greek. The real numbers are, at most, estimates or qualified guesses. Besides, the real numbers reflect both the proportion of native speakers of these languages residing in the USA, and the level of attractiveness of the Internet segment referring to their mother tongue. In 2002, the idea of accessing the Internet in a mother tongue does not seem too attractive to old or new emigrants from Vietnam or Indonesia who reside in the USA, even if some of them might have problems mastering English. Thus, the local segment of the Internet is called "attractive" if it really attracts visitors from outside the local community. To simplify the definion, might call "local" every non-English-based segment of the Internet, however large the proportion is of the world population. Probably not all of these segments are attractive, as tried to show earlier.

Of course, each website, and/or group of closely connected websites, is designed to attract outsiders, and the number of hits, clicks, visits, sessions, or the target audience reached, are very important measures. But the segments based on ethnic languages usually include every type of website which advertises itself, and yet the segment itself has or lacks qualities that might make it attractive. Thus we save the term "attractiveness" for the most broad - based on ethnic languages other than English- segments of the Internet.

The Internet related culture in Russia is very young, it has been developing since 1990 - the year when the Russian computer network established a connection with Western European. The Internet population in Russia is not numerous and the WWW content is modest in its quantity. Chronologically, the first published estimates of the number of networkers in Russia were made by an American and refer to 1994. Globally, actual approximations of the Russian Internet population and the WWW content are as follows.

About 5,000,000 high-quality PCs with modern software are in service in Russia. About the same number of older PCs are in service and are not likely to be connected to the Internet. The Russian segment of the Internet includes approximately 700,000 connected computers and more than 350,000 unique IP-addresses. The computers with access to the Internet are rather innumerous; that is, users most often access from office/school/university, and less often from home. Sociological data confirm that this is true. Statistics of web traffic suggests that weekend traffic is less tense than work-time traffic. This might be partly shadowed by the fact that there are so many time zones in Russia due to its huge area.

There are more than 140,000 domains in Russia, and among those there are more than 75,000 second-level domains. One should take into account that, usually, only .RU and .SU domains are counted. It is known that some Russian sites belong to .ORG, .NET or .COM, but they are probably innumerous, and webcrawlers can hardly identify them. At the same time, some servers within the .RU or .SU area are physically located outside Russia.

The annual increase of Russian-language content is over 200%. In January, 2002, it contained about 60,000,000 indexed documents, which are well rubricated and fully searchable. The global volume of information is estimated as close to 1Tb, and an average web-server contains as much as 2.5Mb information.

Hundreds of Russian-language newsgroups are active; FIDONet users participate in over 2,000 moderated echo-conferences. The most universally used search engines are Rambler and Yandex which were originally Russian productions. Lycos started business in Russia in 2001 and organized its Russian service. Major world-wide search engines, such as Google and AltaVista, operate with Cyrillics.

One need not know Russian or read Cyrillic to search through Russian web content using originally Russian search engines, since the queries are automatically translated. A complicated morphology of the Russian language is handled and sentence parsing is performed. All the advanced products support major alternative Cyrillic charsets.

Estimates of the number of Internet users in Russia are being made by several sociological agencies. The agencies are mostly Russian, but agencies from abroad are active as well. Research data has been available since 1996, when the first representative research was performed. Sadly, since that time, the number of Internet users in Russia has been the subject of strong disagreement among the few agencies that are systematically carrying out fieldwork in the area.

The number of actual Internet users in Russia were 3,300,000 (3.6% of adult population), of whom 60% are men and 40% are female, more than 20% are teenagers, 40% are aged between 20 and 29, about 20% are aged between 30 and 39, and only 6% of Internet users in Russia are over 50 years of age. These data refer to November, 2000. Since that time no reliable data have been published, and it is only reasonable to assume that the number of users have increased and exceed 4,000,000. The Russian Ministry of Communications estimates the number of permanent Internet users as 2,500,000 to 3,000,000 and the number of casual users as about 5,000,000. Residents of Russia comprise approximately 1% of the world-wide Internet population.

Economically and politically, the Internet means a lot to all those who identify themselves with the Russian segment of the Internet. The previous tense links between the residents of the newly independent states that have formed after the collapse of the USSR, show a tendency to be called, at best, virtual. Personal contacts are severely reduced both in quantity and quality due to ever-increasing travel, phone and postal tariffs, customs emerging at previously transparent frontiers, decrease in the exchange of TV programs and in the circulation of newspapers, problems with the exchange of the unequally "weak" national currencies, acts of brutality and/or expectations of hostility towards the representatives of minority ethnic groups, etc. In addition to inefficient personal contacts, and in many ways due to the same reasons, economical and industrial ties have reduced too.

Only political extremists and/or marginal nationalists would deny that the huge territory - one sixth part of the globe surface (with the exception of oceans and seas) - is possibly open for re-establishing traditional forms of links and for maintaining those working and personal contacts that have happily survived - on the new information technologies basis. Global telecommunications seem to emerge just in time to mediate processes of filling in the "inefficient communication" gap.

Telecommunications are heavily used on an organization-to-organization level. But this is not enough. The Internet, mobile phones, videoconferencing systems, etc. bring a new potential to solve the problem of keeping contacts among people who were born and got to know each other in their common country, lived there a life-long period of time, spent leisure time together and paid visits, but recently found themselves residing in diverse states and facing severe problems when they make attempts to engage personal or social contacts.

For the purposes of mediation in establishing contacts and in the formation of distant communities the WWW content is of special importance. The tendency is that Russian web content is widely used by members of Russian language communities outside Russia. Non-residents of Russia are very active in newsgroups, including both ethnic Russians and ethnic non-Russians. The latter speak Russian fluently, they are often born and bred within the Russian culture and usually are educated in Russia and/or in the Russian language. The majority of this special Internet community population found themselves outside Russia when the Soviet Union collapsed in 1991. They now reside in the former Soviet republics - now independent states.

A minority of the community members are permanent or temporary residents of diverse countries (usually, but not exclusively, Germany, the USA, and Israel), and they feel themselves closely connected with the Russian culture. Global Internet Statistics estimates the number of Americans who regularly access the Russian segment of the Internet is 100,000. The migrants represent what is often called the Russian diaspora: they continue to speak Russian, and they support websites with the content devoted exclusively to Russian culture. One of the most important web sites of this type is Little Russia in San Antonio, Texas - it has already attracted interests of communication researchers.

The diaspora consists mostly of "new" migrants, and partly of "old" emigrants from Russia, including those who were born abroad. One should take into account that in the twentieth century there were at least four periods of massive migration from Russia to the West and/or East, and respectively four waves of migrants can be differentiated depending on the time of their escape from the Soviet regime.

The other source is the webservice SpyLog which traces most of the navigation within the Russian segment of the Internet. The SpyLog tracker presents data that approximately 40% of navigations are systematically made from outside Russia. Among the visitors from abroad there are native Russian speakers and those who speak Russian fluently, as well as foreigners meeting certain problems when they read Russian sources. The browser interfaces are in Russian for the majority of navigators from the former Soviet states, with the exception of representatives of the three Baltic States, who use browsers with English or local language interfaces. Almost all visitors from Western Europe, Israel, or the USA use the browsers with an interface in the local language. Thus, browsers give some hint about visitors to the Russian Internet segment and the degree of their closeness towards the Russian Internet culture.

The reasons have been deduced from discussions with colleagues, personal friends, or acquaintances residing outside Russia. These discussions have been held over several years, beginning approximately in 1994. The discussions have never taken place in a laboratory setting, and have never taken the status of a formal interview. No formal procedure of handling the results was used, but for classification. More sophisticated procedures would require discussions to be recorded. The number of respondents are approximately 70 non-Russian residents. The above-mentioned statements should prove that Russian content on the Internet, though incomparable to the English-based content in quantity and quality, is nevertheless one of the attractions of the Net for non-residents. It is noteworthy that modern residents of Russia are about 60% of the overall population of the former USSR. The reasons for former Soviets from outside Russia being attracted with Russian-based content are mainly as follows:

— Lack or shortage of attractive web site content in their ethnic languages. The majority of post-Soviet states (with the single exception, perhaps, of Estonia) are far behind Russia in the rate of people's access to the Internet and in the number of web sites.
— Poor (if any) command of official ethnic languages of post-Soviet states by ethnic Russians residing there.
— Poor literacy skills in their mother-tongues of numerous non-Russians in the post-Soviet states. Often they speak their ethnic languages fluently but have problems with writing and reading, i.e. with composition and comprehension of scholarly texts due to the fact that when acquiring formal education they used Russian solely.

Political and/or religious leaders of some ethnic groups who used Arabic or Mongolian alphabets in the pre-Soviet period feel pressure to return to cultural origins. It seems only natural for some of them to cease using the now-actual alphabetical systems that are based mostly on a slightly modified Cyrillic and to make a change to the historically proven way of writing and reading. These decisions result in a peculiar sociolinguistic situation when different generations speaking the same language might soon have no common literacy to exchange written texts, including private letters. Happily this process is slow and the situation of cross-generational misunderstanding is perhaps a futurist one.

In less populated countries it is impractical to originate and support newsgroups and to compose valuable websites content in ethnic languages, except for purely official information and for data of exclusively local interest. Even if ethnic language newsgroups and websites are available, more diverse views are expressed and more valuable information displayed might be found in Russian-based newsgroups and websites. Usually, in post-Soviet states, educated people have some command of English, but their command of Russian is in most cases much better. This fact explains, at least partly, the reasons for their attraction to the Russian language segment of the Internet. Individuals might feel nostalgia towards older times when all the citizens of newly independent states stayed closely united in the USSR; they maintain interest in news from the former parent state and in contacts with former compatriots.

Media in some post-Soviet states are less independent compared to the Russian media. Residents of these countries may find it convenient and reasonable to get access to less censured news via login to the network versions of Russian news agencies. Usually they would try to know more about internal/external politics in their own states — this is a continuation of an old tradition of Soviet times to learn internal news through media from abroad. Politically-minded people from the former USSR - both within and outside Russia - tend to keep group discussions of numerous political issues; some of them blame the former communist regime, and some of them blame the modern ones.

Some representatives of Russian-speaking diaspora residing outside the territory of the former USSR find it attractive to take part in newsgroups and/or to originate/construct/support websites in the Russian language. Their motivations differ: they give consultations on diverse aspects of foreign life; being competent in web-related issues they share knowledge with subscribers to Russian newsgroups; they consult and support originators of network-related projects in Russia; they try to find people they used to know earlier in order to keep personal contacts; they try to find ways to share their hobbies with those speaking their mother tongue; they originate and support websites containing diverse collections.

Russian Internet experts are often believed to be more advanced, experienced and competent than their colleagues from many other post-Soviet states. The outside-Russia networkers, trying to enlarge their competence in the information technology related news and to follow discussions of the principal hard/software decisions, find it worthy to subscribe to the Russian language newsgroups and to surf the updated reviews published on Russian web-sites. Usually they find in this way much more valuable information compared to the quantity and quality of information available in the ethnic languages of the states in which they reside. These are some of seemingly important reasons which partly explain the attractiveness of the Russian segment of the Internet for the Russian speaking foreigners.

The modes of Internet usage are largely dependent on the level of command of English and on the competence in behavioral patterns peculiar to North American culture. Like many other activities, especially those dependent on the technologies' usage, Internet-based communication and cognition relies on a heavy use of English. Few ethnic languages claim to be attractive for foreigners participating in online discussions and/or visiting websites. It is stated that the Russian segment of the Internet is such a cluster of attraction.

The qualitative and quantitative parameters of the Russian segment of the Internet are described briefly and give evidence that these parameters are developing rapidly. Some probable reasons that might explain the foreigners' interest to Russian-language web sites, discussion lists, or web-forums are proposed. These foreigners are partly ethnic Russians residing in the former Soviet states outside Russia, and partly non-Russians who feel themselves connected to the Russian culture; a much smaller but influential part constitute former Soviets who have emigrated to the Western/Eastern states outside the borders of the former Soviet Union.

## Communication Culture of Cyberspace

Power is a complex notion in social theory and discussion of it will be necessarily short. Here will be taken to mean the various ways in which different individuals have different possible actions they may be able to take.

Such an interpretation is open to a voluntarist reading and a structuralist reading; it is also open to different interpretations of the moral or ethical meaning of power around the two poles of power as the name for that which creates social order and power as a form of domination.

This interpretation of power is derived principally from theories developed by Max Weber, Barry Barnes and Michel Foucault. The legitimacy of this interpretation rests on this matters aim of defining power in cyberspace, not the nature of power itself. Having taken a definition of power from a broad reading of social theory, it is the utility of this definition in helping to define cyberpower that is most important and not an answer to the complex question 'what is power?'

In front of a computer screen, reading the glowing words we confront singularity before building a sense of others in the electronic world. There is a double sense of individuality here. First, people must simply connect to cyberspace by logging in, almost certainly involving the individual entering their online name and their secret, personal password to be rewarded with their little home in cyberspace. The first moment in cyberspace is spent by nearly everyone in his or her own individualised place. Second, moving from this little home to other virtual spaces usually involves further moments of self-definition; for example, choosing an online name, choosing a self-description or outlining a biography. The experience of logging on occurs not only when entering cyberspace but is repeated across cyberspace as we enter name and password again and again. The two key areas where being an individual in cyberspace allows different actions to be taken than in offline life can be called identity fluidity and renovated hierarchies. These will be briefly explored in turn.

Identity fluidity is the process through which online identities are constructed. It remains true that in all sorts of online forums an individual's offline identity cannot be known with any certainty. With the reasonably well-documented instance of a conservative Jewish, teetotal, drug-fearing, low-key, sexually awkward, male, abled, psychiatrist convincingly posing as an atheistic, sexually predatory, dope smoking, hard drinking, flamboyant, female, disabled, neuro-psychologist are in the presence of a potential disconnection between online and offline identities. However, it would be a misconception to draw the conclusion that identity disappears online. Identities that constrain, define and categorise exist online, but these identities are made with different resources to offline identity. Broadly online identities are constructed out of two types of indicators: identifiers and style. Neither of these mandate that someone's offline identity must reappear within their online identity, though there are many ways in which a repressed offline identity may return in the midst of online fantasy.

Identifiers are the addresses, names, self-descriptions and more that designate contributions to cyberspace. Email addresses are the most common form of identifier and can illustrate the resources provided by identifiers for the construction of virtual identities.

The identity of the sender of an email can be constructed, in part, from their email address. All sorts of identifiers abound in cyberspace and there is no restriction on only having one, in fact there is almost a compulsion to creating several. Identifiers include the signatures people place at the bottom of their email, the often lengthy self-descriptions Muds and some discussion groups allow, the various names might choose or have imposed on for various lists, newsgroups and so on and even the visual avatars some are developing for three-dimensional shared virtual places. All these function as long or short, more or less freely chosen names; they are the virtual equivalent of seeing someone's face and being able to think male or female, black or white, old or young, and so on. The second type of resource for identity is style.

Many virtual places have resident celebrities whose styles are instantly recognisable to other participants and anybody who participates repeatedly will eventually come to have their style recognised. Groups also provide certain stylistic resources. Abbreviations are common in the typed world of online discussions, such as 'btw' for 'by the way' and groups sometimes generate their own specific abbreviations.

Online personalities are constructed through resources that are different to offline, preventing the use of many offline tactics for identifying identity. Online characters are constructed and judged through a number of markers that replace offline ones; addresses, handles, signatures, self-portraits and styles. Identity is both present in cyberspace and is different to non-virtual space, the only mistake here would be to assume that the powers the flow around offline identities-such as those around gender or race-are absent online, instead of identifying the particular forms of identity which exist in cyberspace and on which power takes hold.

Renovated hierarchies are the processes through which offline hierarchies are reinvented online, with many online resources undermining offline hierarchies while also defining new hierarchies. The second component of online life that appears obvious to the individual online is that it seems anti-hierarchical. Attempts to censor or restrict access to parts of cyberspace can often be by-passed.

Further, communication from many people to many people is close to the norm in cyberspace. This opens participation in decision-making, creating the potential for conclusions to be reached in more egalitarian ways than offline. Three ways in which hierarchies are affected will be noted; identity, many-to-many communication and anti-censorship.

Identities are one of the key building blocks in offline hierarchies, but if no-one knows are black or disabled online then cannot be placed in a hierarchy on that basis. This seems undoubtedly true, and to the extent that someone can keep their offline identity separate from their online then offline hierarchies based on identity can be dislocated. However, identity does not disappear online but is remade according to the rules of identifiers and styles.

This means that specifically online hierarchies can be expected, such as that noted by Branwyn who was told be an online sex enthusiast that 'In compu-sex, being able to type fast or write well is equivalent to having great legs or a tight butt in the real world.' All the various resources available for the construction of online identities will also function to create online hierarchies. Someone's witty and knowledgeable posts to a newsgroup, their style, may mean their claims are treated more seriously than a newcomer and can be reasonably certain that many may treat an email from billg as being of higher importance than many other emails we receive.

The second way hierarchies are dislocated is through many-to-many communication and its ability to include people in decision making. The inclusion of people in offline decision making is limited by the need to meet together, to only speak one at a time, to overcome the hierarchies of identity and so on. Offline hierarchies can be undermined through the broader access to information that cyberspace offers. In the UK in 1999-2000 there has been a discuss over the ability of patients to research their illness and treatment over the Internet and whether this is undermining the authority of physicians. Without entering into this particular debate, can note that it is one sign of the way the hoarding of information as a means of generating a more powerful position in society, particularly by professional bodies, can be undermined by cyberspatial communication.

Offline hierarchies are undermined in online life is by the censorship evading properties of the Internet. Not only is there a greater pool of expertise available but information that governments or courts might have restricted is almost impossible to hold back once it is free in cyberspace. The global nature of cyberspace is important here, as it only requires one country connected to the net to allow the publication of some information for that information to be let loose in cyberspace. Information restricted in an offline nation-state will then be available in cyberspace, subverting the national boundaries that have helped in the past to control access to information. A global informational space undermines regional or national attempts to restrict access to information.

While it is untrue to say that hierarchies are absent in cyberspace, it is also true to say that hierarchies there are made according to different rules than online hierarchies - rules that at their most hopeful make someone's ability to write creatively and knowledgeably the basis of higher positions in a hierarchy - and that many offline hierarchies are undermined by cyberspace's powers. Taken together identity fluidity and renovated hierarchies can also be seen to rely on the third component of cyberspace from the individual's viewpoint, because both rely on cyberspace's nature as an informational space. Sharing information allows the construction of identities; self-descriptions, signatures, styles are all constructed out of the words that pass between people. The renovation of hierarchies in offline life by cyberspatial communication results from different methods of sharing information and different access to expertise in cyberspace. When individuals experience cyberspace they come to the third recognition that life in cyberspace is fundamentally constituted by information.

If identity fluidity, renovated hierarchies and informational spaces constitute cyberpower from the viewpoint of the individual then power at this level must be understood as the possession of individuals, who can utilise the various abilities offered by these three to impose their will. If we pause and reflect on what it seems that cyberspace and the Internet offer as individuals, then the ability to remake our identity and to renovate the hierarchies are caught within make cyberspace appear as a place that offers various powers. These powers can used by individuals to take various actions they had not previously been able to; whether it is the desire of a woman to have gay sex with men or of a parent to understand the treatment their doctor is recommending for their child.

From this conclusion can understand the enormous hopes and commitment cyberspace sometimes draws from people, because viewed as a space based on individuals the main effects of cyberspace seem to be to offer various powers to act. It is from this perspective that the most hopeful visions of cyberspace derive. It is also a perspective based on the repeated and ongoing experience everyone has that entering cyberspace marks us as an individual, it is not a perspective that is simply naïve but it is one reinforced by the daily experience of millions who have virtual lives. Cyberspace is here understood as the land of empowerment of individuals, of reinventing identities out of thought.

Many people report a transformation, often slow, in their perception of online life. From an initial combination of bewilderment, glee and skepticism many come to accept the online world as normal - from MUD dragons to email. With stable online identities, in whatever forum they exist, people begin to have ongoing conversations, to meet the same people and learn their peculiarities. The particular rules of different corners of cyberspace become clear and normal, but then it is often realised that the individual is no longer the final cause of online life, for communities have emerged. The transformation is not magical but sociological. Even communities that begin by assuming the sovereign individual is primary soon come to realise that collective responsibilities and rules appear, created by many and over which no one person has control.

The idea of one person constituting a language or creating a society is strictly speaking absurd; anyone can invent a word but to have it understood means having a community. This transformation to seeing cyberspace as inhabited by collective bodies is not a simple opposition between individuals and collectives but an inversion of the relationship between these two. Individuals possessing cyberpowers can produce collective bodies, but individuals are here understood as being the fundamental constituent of community in cyberspace. What is often realised in contradiction to this individualism is that collectives may create the conditions under which certain forms of individuality can be realised.

The collective becomes the fundamental cause. At both levels of cyberpower both virtual individuals and virtual communities exist, but their relations are reversed. Cyberpower of the social derives from the belief that individuals have their possible actions defined by the collective bodies they are part of.

All of who use the Internet and enter cyberspace rely on a range of technologies. Any action can take in cyberspace, from changing a gender in a Mud to buying and selling stock, can occur only because we have entered an electronic space created and maintained by various technologies - IP, routers, personal computers, optic fibres, modems and so on. Our individual powers in cyberspace will be defined by the technology we are using and the capabilities this technology offers. The fundamental realisation that cyberspace is not just about powers individuals can use but also about the things that create those powers for all users, is essentially the same realisation as that there are collective bodies in cyberspace that both create and restrain the nature of individuality in cyberspace. This is because the communities that provide the basis for virtual individuals are essentially constituted by technologies. For example, the conditions for Usenet communities are primarily defined by the fact that Usenet technologies create discussion group made out of posts.

The nature of individuality in Usenet is constrained by the nature of its discussion group software and hardware. Different forms of individuality are possible in Muds, such as building virtual homes, and on the Web, such as use of graphics. All these powers that cyberspace offers the individual are based on communally created and experienced technologies.

If we assume cyberspace is the realm of societies and collectives, then a form of technopower becomes visible. Technopower is the constant shifting between objects that appear as neutral things - keyboards, monitors, email programmes - and the social or ethical values embedded in these objects by their designers and producers. Each questions the other. If email software allows many-to-many communication, we can ask why?, who made the software do this?, what results come from this? and in asking we may open up the inhumane appearance of the programme to find humans who embedded their ethics or ideals in a programme.

Technopower underpins the social structures of cyberspace through a constant shape-shifting between seemingly inert technology and seemingly alive values. This does not mean that every piece of technology was created with a form of technopower in mind. Not every technological thing we encounter or rely on has been purposefully designed to constitute the social life it actually constitutes. The unexpected result is always possible, like the emergence of teflon from the space race. But what will be found behind each thing will be humans in social spaces, making decisions within institutional and technological contexts. Dead technology always opens on the living, just as it is the living who create technology.

Technopower is constituted like an infinite series of Chinese boxes, each opening onto another little model of itself, and each layer composed of the same elements, inert seeming technology and alive seeming values. Technopower can also be seen in offline life; from car engines to ice cream, we live surrounded by technological artefacts that leak social values from every crevice. The difference between online and offline here is that online social forms are constituted fundamentally, if not totally, from technopowers. When we adopt the perspective of the social in cyberspace, we lose sight of individuals and their powers and bring into focus these impersonal technopowers that constitute the very possibility of cyberspace in the first place.

At the level of the social, cyberpower is a technopower. However there is more to say, as a particular direction to technopower can be defined. Technopower in cyberspace is governed by the ever increasing reliance by users on technological tools, that time after time appear as neutrally pointing the way to greater control over information but time after time result in different forms of information constituted by the values inherent in the new tools.

Information is endless in cyberspace and creates an abstract need for control of information that will never be satisfied. The direction of technopower in cyberspace is toward greater elaboration of technological tools to more people who have less ability to understand the nature of those tools. Control of the possibilities for life in cyberspace is delivered, through this spiral, to those with expertise in the increasingly complex software and hardware needed to constitute the tools that allow individual users to create lives and societies.

What can be called the technopower spiral is constituted out of three moments. First, there is the ongoing and repeated sensation of information overload in cyberspace. Cyberspace is the most extreme example of a general acceleration in the production and circulation of information. For example, cyberspace encourages people to produce more information rather than passively consume it. Information moves faster and in greater quantities in cyberspace than in other space.

Most powerfully, cyberspace increases information by releasing it from material manifestations that restrict its flow and increase its price. Ideas embodied in books have inherent costs and restrictions on the number that can be produced and the speed at which different people can obtain them. Information is largely freed of its material form in cyberspace. This constant increase in the sheer amount and speeding up of information leads to the experience of information overload. While the notion of having too much information might seem paradoxical, it is also the case that only a certain amount of information can be dealt with at one time. As early as 1985, Hiltz and Turroff estimated that computer-mediated communication resulted in what they call superconnectivity, whereby individuals' connections to each other increase ten-fold and this is a result reinforced by recent work.

Too much information or information too poorly organised leads to information overload. How many of know the feeling of signing up to an email list and then finding the constant flow of emails means messages have to be deleted before being read and the group resigned from? How many of search the Web for a particular topic only to end up with megabytes of files or piles of print-outs destined never to be read because there is simply too much? Cyberspace increases information's velocity and size to such an extent that information overload is a constant experience of virtual lives.

The second moment in the spiral in technopower is the attempt to master whichever moment of information overload has occurred. This can be done simply by switching off, but doing so removes all the powers the individual might feel they benefit from. Instead, information overload is constantly addressed with new technologies. Various solutions to the glut of information that cyberspace produces have been created, from news services that email news bulletins once a day to ticker tapes that produce a constant flow of stock prices or news flashes across a browser. Maes lists intelligent agents that schedule meetings, filter Usenet news and recommend books, music and other entertainment. All these share a number of traits. First, they interpose some moment of technology between user and information. This is always simultaneously a moment in which technopower is manifested and articulated because some technological tool, appearing as a thing yet operating according to values, is the method of controlling information overload. Second, the devices themselves produce information problems because they need to be installed and used properly. No matter how sophisticated such a device is the user will need to understand how to manage the device or risk being controlled by it.

Third, new tools nearly always make more information available and cyberspace easier to use, tending to create a new overload. This seems too paradoxical to be true, as the goal of many tools is to reduce the amount of information received by focusing or managing it in some automated way. However, the very success of any such tool tends toward the production of more information because it makes gaining information more efficient and there is always more relevant information waiting out there in the infinite reaches of cyberspace. Problems can be expected to re-emerge with the devices that have become essential to information management themselves producing too much information.

The technopower spiral is completed and re-initiated with the emegence of a new problem of information overload. This spiral of overload, tools, more overload and more tools is fundamental to technopower in cyberspace. It means that as individuals pursue their powers in cyberspace, they constantly demand more technological tools to master the seemingly infinite amount of information at their disposal. Technopower is constantly elaborated to meet the demand to control and manage information in cyberspace, thereby ensuring cyberspace constantly becomes more and more technologically complex. This, in turn, means that the ability to act in cyberspace is constantly elaborated by those who have technological expertise, either personally in figures such as hackers or by managing those who have expertise, as Bill Gates or Linus Torvalds do.

Accordingly, cyberpower of the social is a power of domination, through which an elite based on expertise in the technologies that create cyberspace increasingly gain freedom of action, while individual users increasingly rely on forms of technology they have less and less chance of controlling. If the cyberpower of the individual was a hopeful form of power, pointing as it does to the increasing range of actions cyberspace can help an individual to take, then cyberpower of the social is pessimistic because it reveals networks of interactions that increase the ability to act of an expertise-based elite.

Cyberpower encompasses both the perception that cyberspace offers power to the individual and that societies are defined by an expertise-based elite whose class power is ever-increasing. To complete this mapping of cyberpower, a third level needs to be outlined, for cyberspace exists not just in virtual lives and communities but also in dreams.

People are part of imaginary relationships that define societies and nations. A nation can be thought of as 'an imagined political community-and imagined as both inherently limited and sovereign'. It is imagined because it is impossible for all members of the community to meet, they must hypothesise their commonality. It is limited because there are always borders and beyond those borders there are other nations. Finally, it is a community because, regardless of actual inequalities between members of a nation, it is always conceived as a 'deep, horizontal comradeship' in which all are equal as members of the nation. A further recurrent characteristic of imaginaries is that they offer hopes and fears that appear as real projects just one or two steps away from completion.

Much of the urgency people draw from imaginaries stems from this sense of being nearly but not quite completed, meaning people feel a need to act quickly to either prevent the imagined disaster or bring on the imagined benefit. A similar imagined community exists in cyberspace and cyberspace's imaginary is driven by this compelling feeling that change is near. The content of cyberspace's imaginary is structured by a twinned utopia and dystopia that both stem from the claim that everything is controlled by information codes that can be manipulated, transmitted and recombined through cyberspace.

On the one hand, cyberspace gives rise to many hopes, including or especially the oldest human fantasy of immortality. Certain people believe that within our lifetime we will be able to separate our 'consciousness' from our body and upload it onto silicon. If the essence of our self can be made virtual, then what better home than cyberspace, where all the immortals would be able to meet.

In a related way, some believe cyberspace is becoming the one organic mind that will constitute a higher consciousness akin to god; a hope that has haunted the hippy mind at least since John Lennon declared everyone to be a god. On a less metaphysical plane, many see the cyberspace-based cyborg as the figure who will lead beyond the oppressions of this world - gender, class, race, sexuality and other - because the cyborg transgresses the oppositions between nature/culture, human/animal, and man/machine and in doing so helps to destroy the binary oppositions oppression is based on. Whether the heaven is a selfish one of personal immortality, a spiritual one of god-like hive mind or a materialist one of post-revolutionary utopia, cyberspace offers the possibility that this heaven is coming into existence.

On the other hand, cyberspace gives rise to many fears. Cyberspace is seen by many as the ultimate tool of surveillance. If repressive power operates according to the principle 'visibility is a trap', then cyberspace may make even our smallest interactions visible. A world is foreseen in which all interactions with the government, in education, in shops or any other institution will be carried out by electronic means and cyberspace will allow all these interactions to be linked, collated and examined. People will be considered guilty until proven otherwise because all interactions will be examined to see if they meet the expected pattern and if that pattern is not met then an investigation will be triggered.

More insidiously, information will be collected from everyone in the guise of serving their interests, such as with supermarket reward cards that offer special deals or money back in exchange for connecting the nature of our purchases to our socio-economic identity. We might also be implanted or tagged, cameras on public streets now constantly survey facial heat images looking for a wanted suspect and the USA and UK governments currently record and examine all fax, telephone and Internet communication in Europe and elsewhere.

All these nightmarish possibilities can be made real because cyberspace provides the perfect medium for collecting and transferring the phenomenal amounts of information that are necessary to this hellish vision of the totally supervised society. Both radical hopes and fears in cyberspace result from a belief that everything can now be manipulated through information codes. We may become immortal by turning our 'self' into an information code that can live in cyberspace or we may become prisoners in an open society because our 'self' can be defined by the information cyberspace draws together.

Certain particular visions are collectively imagined from the realisation that everything, even life itself in DNA, is an information code. This form of power operates not, as might seem obvious, to create or resist the imagined heavens and hells but by providing some of the unifying thoughts that allow individuals in cyberspace to recognise each other as members of the same community.

As different people come to grasp some parts of both sides of the imaginary, they come to recognise themselves as part of something larger than the individuals they meet or communities they participate in. Whereas cyberpower at the level of the individual offers possessions that enable people to act more widely and cyberpower at the social is a network that creates increasing power for a techno-elite, cyberpower at the imaginary constitutes the broad social order of cyberspace by providing dreams and nightmares through which individuals and communities come to recognise they are part of something greater than themselves.

Power is the condition and limit of politics, culture and authority. Cyberpower aims not at the immediately obvious forms of politics, culture and authority that course through cyberspace but at the structures that condition and limit these. A certain complex form of power that operates on the three levels of the individual, the social and the imaginary now careers through the virtual lands, directing conflict and consensus towards certain distinctive issues and social structures.

No one level of cyberpower determines or dominates the others. In particular, the powers of the individual and the social are in constant battle. The powers the individual gains in cyberspace, such as cryptography, may contradict the domination of a technical elite, just as the technopower spiral may lead individuals to increasing reliance on technological tools whose metaphorical bonnet they cannot hope to open. We can expect these two levels to swing back and forth, with individuals gaining powers against an elite only to find they have given birth to another part of the elite.

Informational libertarianism or anarchism gains its special place as the political discourse of cyberspace here because it emphasises both individual liberty, speaking to individuals and their powers, and that cyberspace produces the best possible outcomes through free markets, speaking to the elite as a justification for their growing control.

Libertarianism on the net has at its core a doubly articulated concept that connects individual liberty to free markets, allowing the one ideology to speak to both the elite and the grassroots. This does not mean libertarianism, or the sort of informational anarchism that permeates hacking, will be universally celebrated on the net but that it is a uniquely equipped ideology through which politics on the net will be played out. This analysis does not endorse libertarianism, it tries to identify why variants of libertarianism and anarchism have formed such a powerful political language in cyberpolitics. And many politically motivated hackers will be found endorsing sentiments of freedom of information flows in anarchistic language.

Cyberpower points not to the ultimate dominance of elites, though it clearly identifies the burgeoning power of an elite, nor does it predict the ideal of individual empowerment, though it makes conspicuous the ongoing creation of powers for individuals in cyberspace. Cyberpower points to these processes continuing, driven by dreams and nightmares. When examining cyberpower we must always be aware of the roar of battle and the complex conflicts that define virtual lives, elites and dreams.

## References

David Koepsell, (2000). *The Ontology of Cyberspace*, Chicago: Open Court.

Irvine, Martin. (2006). "Postmodern Science Fiction and Cyberpunk", retrieved 2006-07-19.

Oliver Grau: (2003). *Virtual Art. From Illusion to Immersion*, MIT-Press, Cambridge.

Sterling, Bruce. (1992). *The Hacker Crackdown: Law and Disorder On the Electronic Frontier*. Spectra Books.

William Gibson. (2004). *Neuromancer:20th Anniversary Edition*. New York:Ace Books.

Zhai, Philip. (1998). *Get Real: A Philosophical Adventure in Virtual Reality*. New York: Rowman & Littlefield Publishers.

# 6

# Nature of Information Economy

Information economy is an economy with an increased emphasis on informational activities and information industry.Manuel Castells states that information economy is not mutually exclusive with manufacturing economy.He finds that some countries such as Germany and Japan exhibit the informatization of manufacturing processes. In a typical conceptualization, however, information economy is considered a "stage" or "phase" of an economy, coming after stages of hunting, agriculture, and manufacturing. This conceptualization can be widely observed regarding information society, a closely related but wider concept.

There are numerous characterizations of the transformations some economies have undergone. Service economy, high-tech economy, late-capitalism, post-fordism, and global economy are among the most frequently used terms, having some overlaps and contradictions among themselves. Closer terms to information economy would include knowledge economy and post-industrial economy.

## Conceptualizing the Information Economy

Three processes related to the information economy have been unfolding concurrently, each according to its own dynamics and speed. A series of alternative conceptualizations of the information economy has emerged. The subfield of the economics of information is coalescing out of the merging and expansion of several strands of work in the literatures of economics. And empirical developments - including the appearance of new types of organizational form, shifting market activities, and the sustained production and distribution of goods and services - continue in their own multiple ways, unaware of, and unconcerned about the efforts of scholars to contain them.'

The three are, of course, not unrelated. Conceptualizations of the information economy inform the making of policy and interact with the definition, bounding, and attitude of the nascent economics of information. Developments in the economics of information - most profoundly explored by Don Lamberton - expand the kit of tools available to analysts and decision-makers in both the public and private sectors. The policies which result are among the structural forces shaping the environment in which economic activities unfold. Thus, these elements are mutually constitutive.

The global economy is undergoing a shift from dominance by market relations to dominance by relations within organizations and other forms of networks. This is happening just as innovation in the information infrastructure enables evolution of organizational form itself. The simultaneity of various processes means there is more than one thing going on at a time. This is particularly so when one looks across levels of the social structure or uses different geographical or cultural loci as lenses. Sheer turbulence further adds to the complexity of economic activity.

In coping with these changes and complexities pragmatically and analytically, enough difficulties have been experienced that a paradigm shift in economic thinking appears to be required. Thomas Kuhn argued that the shifts in the basic paradigms of human understandings about the nature of the world that are the history of science come about as the result of a process that begins with the accumulation of empirical evidence that does not fit into the existing paradigm. Efforts by scholars to fit conflicting evidence into the existing paradigm result in ever more complex models; the extraordinarily elaborate astronomic models of several hundred years ago provide a vivid example. However, as the counter-evidence builds, it becomes ever more difficult to incorporate it into any model based on the existing paradigm, no matter how complex. Meanwhile, other thinkers, scholars, and scientists will have begun suggesting alternative paradigms. Ultimately, the complexity of models, in combination with the amount of evidence that will not fit within them, causes the existing paradigm to fall and an alternative paradigm to come to dominate thought and scientific research in what Kuhn referred to as a 'scientific revolution'. The period during which alternative paradigms are being discussed can be quite turbulent.

The three ways of conceptualizing the information economy that have appeared represent different stages in the process of paradigm change, from denial and resistance, to acceptance and experimentation, to tentative first identification of the seeds of a new paradigm. These three approaches differ in the questions they ask about the economy; the assumptions upon which they are based and the theories used to work from those assumptions; the ways in which they respond to the problems encountered during efforts to use neoclassical tools in the analysis of information creation, processing, flows, and use; and their implications for decision-making. The approaches are not mutually exclusive; it is argued here that a variation on the third approach, an enriched network economics perspective, has the greatest validity and the greatest utility as a foundation for further research and for use in decision-making purposes.

The first approach focuses on changes in the nature of information products, and defines the information economy as that in which products in the information sector play a proportionately larger role than they have during other periods of history. The second focuses on changes in the domain of the economy, and defines the information economy as that in which the economy has expanded through commodification of forms of information and its flows never before commodified. The third focuses on changes in the way in which the economy functions, defining the information economy is that in which the market has been replaced by harmonized information flows as the key coordinating mechanism.

The clarification of the different types of conceptualizations of the information economy offered in this chapter may contribute to the emergence of the field of the economics of information. Within the rubric of an enriched network economics approach there is room for many, if not all, of the strands in the history of economic thought which have dealt with aspects of the economics of information - transaction costs, uncertainty, asymmetry, and so on. A mapping of the latter onto the former should identify lacunae as well as complementarities and contradictions, and thus present a research agenda. The chapter begins with a synthetic review of problems faced by neoclassical economic treatment of information creation, processing, flows, and use.

## Problems in Information Economics

The problems described here have all been confronted by economists, businesses, and other entities struggling to deal with information in its various forms, and are popularly discussed as particularly troublesome features of the current economic environment. They are not all exclusively problems raised by dealing with information; many also occur when trying to cope analytically with things that are tangible as well. Here, the focus is on problems which confound efforts to use neoclassical analytical tools in the analysis of information creation, processing, flows, and use.

### The Problem of Creation

From the perspective of neoclassical economics, things are created only in response to demands of the market. The producer and the consumer are distinct from each other, the producer is identifiable, and there are clear distinctions among intermediate product, final products, and capital. Yet, there are several types of information creation out of motivations radically different from those of the market, including those which build and sustain interpersonal and communal relations, symbolic communication, ritual communications, artistic creations, and physiologically stimulated expressions. Acknowledgment of non-market driven information may suggest a positive mandate to ensure that all members of the community have opportunities for all kinds of expression as part of the body of communicative rights. While historically the consumer has not been considered by economists to be part of the production process, there are multiple roles for consumers in the production of information products and services, from that of shaping products interactively (as with a database, or via collaborative work projects) to their role as products in themselves (as pointed out in Smythe's famous insight that viewers are the most important product in the television industry), to their role in crafting the meaning, and therefore value, of a product.

Joint production is increasingly a problem as new work arrangements evolve in the manufacture of material goods, but it is almost always a problem in the production of information services and goods, and is related to the problem of provenance discussed below. Since the relative import of any individual's contribution is impossible to measure quantitatively, the distribution of the profits and other benefits of the production of services and goods is always a political matter. Some information never enters the market as a commodity because it is retained for use by the information producer. This is a particularly fascinating problem in a period of increasing awareness on the part of organizations that determination as to which is to be retained as a form of capital and which information is to be sold as a commodity is one that can be a deliberate choice in this environment.

**The Problem of Time**

Economists working in the neoclassical tradition are concerned about information only when it enters into and as it affects the market, which requires a gap of time between production and consumption that is identifiable and fixed. A number of problems arise because of the fluidity of the relationships between information and time. Huber would elevate problems of time to the stature granted those of space, noting that historically we have been conceptually biased towards questions of space (tele-communications) rather than those of time (chrono-communications). The production and consumption of many informational goods and services occur simultaneously, so that there is no time during which market relations might appear.

Electronic networking enables constant production processes as electronic communities together create information, work collaboratively on texts or other projects, or participate in ongoing conversations. In such an environment, it is difficult to mark the single point at which a product is a commodity, for commodities are by definition stable in form across time and space (this is a particularly interesting problem when dealing with electronic art from the perspectives of collection and dealing). Similarly, there is no final consumption, only accumulation, dissipation, or transformation. In the domain of information, it is next to impossible to mark the time bounds of a transaction. We respond to ideas raised thousands of years ago, conversing with thinkers across time.

From acknowledgment of the social construction of reality by the social psychologists of the 1920s and 1930s, through the refinement and multiplication of the argument by those in cultural studies more recently, to Posner's articulation of what this means for locating and implementing intellectual property rights, the inability to track precisely and accurately the provenance of information has had consequences for efforts to assign property rights and determine the value of various kinds of processing activities. Information is often said to be perishable, an economic issue as it affects the persistence of value over time; stock market information, for example, rapidly diminishes in value to investors over very short periods of time. The locus of perceived value can move, however, from one category of user (the investor or broker) to another (the analyst, and then the historian). Related issues emerge from the inability to store or stockpile many types of information, or the occasional inutility of doing so when it is possible.

Economists dealing with other utilities, such as electricity, are able to analyze statistics about habits of usage and to incorporate the results in projections for planning purposes. With information creation and flows, however, it has to date been impossible to determine regular peak periods of usage. Information received at different points in time differs in value because it is affected by information previously received. The value of information is cumulative over time and the reception of each piece of information affects the value of information that follows. In an extreme example of this problem, Boulding notes that information about the economic system itself changes the system. Thus, information not only flows within an existing economic structure, but also plays a constitutive role in producing, reproducing, or changing that structure. Boulding describes this problem by calling information a generalized Heisenberg uncertainty principle.

### The Problem of Space

In the 1980s, space as an analytical category began to be important across the social sciences in response to the experience of globalization. It appears in discussions about the economics of information because today's information infrastructure permits distributed production, makes space disappear into irrelevance in many instances, and has caused the customer, at times, to disappear altogether. With the growth of trade in services and foreign direct investment (FDI), the problem of locating just where informational transactions take place has become increasingly pressing. With data processing, for example, a transnational corporation headquartered in one country may have computers in another which process data from a third country in response to requests from a fourth. In such a situation, it is not clear whether the transaction takes place in the country of consumption, of processing, of the data source, of the transnational corporate home, or in the telecommunications network itself- an issue which has a number of economic ramifications and is at the crux of the decreasing ability of nation-states to control the flow of money across their borders.

In a sense space can be said to disappear as an economic factor in the net environment, since services can be distributed globally, across political and geographic borders. Some services are altogether non-transportable because they are dependent on some dimension or dimensions of a specific locale. The degree to which this is true in a particular case depends on the service under discussion and the specific part of the world, socio-economic class, and culture of concern. Furthermore, many services can be delivered without the customer being present in the sense of exercising transaction-specific discretion.

### The Problem of Tangibility

The single most salient feature of information distinguishing it for purposes of economic analysis, is its intangibility. The most popular definition of 'services' in economic and trade discussions has turned out to be that offered by the *Economist*: "Anything that can be bought and sold but which cannot be dropped on your foot". The issues unleashed by intangibles have long been acknowledged in the history of economic thought for they infect efforts to deal with the tangible as well; the classic diamond-water paradox expresses this. In the information economy, however, problems in dealing with intangibles come to the fore because it is the domain currently being colonized for economic purposes. As such, it is the theoretical, computational, institutional, and legal Wild West, where rules are yet to be made. In this world, the relationships between information and the material are several.

One of the most widely recognized problems in the economics of information is the difficulty of distinguishing between the value of information and the value of the materials with which it is associated or in which it is embedded. As Babe notes, most economic analysis of information deals with its tangible packaging, not with the information itself. Viewed from the perspective of a single transaction, a text of such inestimable value as the Bible can be bought quite cheaply, while a CD of an ephemeral heavy metal music group may cost quite a bit more. There are those book collectors who never open their books. Unbundling the value of information from its material manifestation or container is one of the most difficult problems economists face. This is true at the aggregate level as well as in the valuation of specific pieces of information, for statistics generally fail to distinguish the non-informational activities of those firms deemed to be in the information sector.

Labeling is a third potential source of value, in addition to the packaging and the information itself. Because labels offer another tier of self-reflexivity, they make yet more complex the relationship between the material and the immaterial, and can, in themselves, trigger the sense of hyper-reality. While the physical environment is perceived in terms of the symbolic, the symbolic in turn acts upon and shapes the physical. Thus, information is not only often embedded in the physical, but also shapes the materials which embody it. This self-reflexivity, too, confounds efforts to develop adequate tools for the economic analysis of information creation, processing, flows, and use.

Problems raised by features that flow from the intangibility of information are, of course, the most difficult of all. Because relationships between information and the material world are flexible and loosely coupled, information is epiphenomenal. It derives from the organization of the material world. As Babe notes, the same material phenomenon may produce different types of information, and different material phenomena may produce the same information. Many types of information never reach a material form, being embedded instead in relations. Efforts to quantify this type of information include the development of accounting systems to deal with intellectual, social, and cultural capital.

Among the most important of these relationships are those which store the information held by a community and pass it on over time. Via an extremely circuitous route, the economic value of traditional communal knowledge is finally coming to be recognized. As transnational corporations fight over the remaining genetic information held in the tropical centers of biodiversity (currently being rapidly destroyed), they realize that it is not enough to capture the plant and animal germplasm. They also need the knowledge of those in the traditional cultures of these places to access and use the genetic information. If one accepts the premise that value is added each time information is processed, story cycles of great antiquity should be recognized as the wealth that they are. The more information available about a particular material good, the more valuable that material good is. Similarly, the more service available for a particular material good, the more valuable it is. Thus the value of tangible goods can be managed by adjusting their intangible features.

## The Problem of Heterogeneity

Heterogeneity is a problem throughout the field of economics, characterizing capital and money themselves. Heterogeneity of form, value, and function of information are also problematic. No one definition can cover all types of information or all of the industries in the information sector. At the same time, single information commodities may have multiple manifestations, unacceptable in an economics that requires specifiability, locatability, and stability of form of commodities across space and time. The constant development of information products and processes further contributes to their heterogeneity. In the information industries, in particular, there does not appear to be a moment of product maturation of much duration.

Any single information product or process is simultaneously valued differently by different people. This feature is shared with the valuing of material goods, but seems particularly critical in the measurement of the value of information. For information more than for other goods, the value in use may be utterly unrelated to the exchange value. For the same person, the same information may hold different value under different conditions: private and public; depending on the speaker; and as constrained by the time, place, and manner restrictions that are accepted, even under First Amendment law in the United States.

The same informational product or process can simultaneously serve multiple functions in the economy. An accounting system, for example, is not only a commodity in its own right; it is also a structural device used in the production of other commodities, providing coordination within and between organizations. It serves related functions, a perspective implicit in efforts by transnational corporations in favor of including trade in accounting services under the rules of international trade.

## The Problem of Inextricability

Perceiving information solely as a commodity in neoclassical terms makes invisible many of the most important consequences of information exchange, such as creation of a public space and the exercise of power. Information is completely dependent upon the social, cultural, political, and ecological elements of its context, treated only as externalities in neoclassical economics. There are two ways to respond to this problem. First, one can turn to non-economic modes of analysis, as Tribe argues when he rejects the use of cost-benefit analysis at the level of constitutional analysis. Tribe points out that cost-benefit analyses - the very stuff of economic analysis of communication policy problems - by definition deal with fixed categories and relations within and between categories, while the mission of constitutional adjudication is to reconsider the categories used and activities permitted within and between them. All communication policy issues, Tribe claims, are constitutional issues.

A second response to this problem is to try to develop alternative decision-making procedures which can handle types of values not previously incorporated in economic analysis. Three types of alternative procedures might be developed: (1) those which identify moments in which quantitative procedures should be abandoned in favor of the use of qualitative procedures appropriate to the particular problem or issue encountered; (2) those which find ways of quantifying variables previously treated as non-quantifiable, so that they can be included in decision-making calculi; and (3) those which offer qualitative decision-making procedures that can be used by working policy-makers in a timely, affordable, and accessible manner.

**The Problem of Inappropriability**

Problems in the appropriability of information are critical, for without appropriation there are no commodities. The fury of concern over intellectual property rights, the vehicle through which property rights are asserted for many (though not all) types of intangibles, reveals how serious a problem this is.

There are two difficulties that arise from the duality of the nature of information as both a private and a public good. There is a fundamental contradiction in treating information economically since the material package in which information is embedded is a private good, while the information itself is a public good. The two meanings of the term 'public good' also sometimes come into sonflict. The common sense meaning looks at public goods as things that should rightfully be accessible to every member of the public, such as water. A public good in economists' terms refers to something whose use does not deprive others of its use. To economists, information lias a public good aspect, for knowledge by one person of something does not deny others of that knowledge. Efforts to expand the ways in which information can be turned into private property mark the hottest battles of the decade.

While the sale of material goods entails the transfer of an object from seller to buyer, in the case of information the seller retains the information, and its use, even after the sale. This is a radical shift in what is understood to be appropriation. Questions about the line between that which is withheld from the market altogether and that which is offered as a commodity, is noted above, are increasingly questions about organizational form and the development of network relationships. When transferring information it is very difficult to restrict use of the information to the buyer alone. Information is, in this sense, said to be 'leaky', for it is very easy for non-purchasers also to use or enjoy the information. As reproduction and transmission costs go down, this problem increases.

**The Problem of Indivisibility**

Agreements about the measurements used to divide objects and groups of objects in the material world have been achieved only recently in human history, contributing to a great spurt in economic activity. Nothing like such agreement has been achieved for 'things' in the intangible world; endless arguments over whether something is or is not 'novel' for intellectual property purposes is one example of arguments over just such measurement issues.

It is difficult to identify discrete pieces of information. In the case of a journal article, is it the entire article, a paragraph, a sentence, an idea, or the stream of literature in which it is embedded? It is not even clear who has the responsibility for unitizing. In the case of a novel, for example, is the writer, the reader (or the filmmaker) responsible for unitizing? In many cases, multiple types of units may be concurrently applicable; which should be deemed most pertinent for economic purposes? Umberto Eco , in a series of essays on fiction, talks about the problem of time in novels, distinguishing between story time (the fictional time span over which the story takes place), discourse time (the time period to which the novel speaks), and reading time (the time it takes to read the book).

Partial information is often useless or even damaging, yet it can be difficult to know whether one has complete information. The persuasive industries, from propaganda to public relations, have found ways to capitalize on this. There are many goods for which sales can be made over and over, such as food and clothing. With information, however, one is often satiated with the first unit of certain types of information, so that repeat sales are not available. Multiple copies are useless unless they can be sold; hoarding behaviors are generally meaningless unless storage of information is the niche in the information economy specifically being filled.

### The Problem of Self-reflexiviry

Both Stiglitz and Arrow have pointed to the problem that arises when purchasers attempt to determine the value of an informational product in order to make a buying decision. Stiglitz described this as a problem of infinite regress, for it is impossible to determine whether it is worthwhile to obtain information about whether it is worthwhile to obtain information, and so on. Arrow, working within a tradition of research into the costs of acquiring information, a tradition in which Coase seminally described the firm as an effort to reduce the costs of acquiring information, points out that the value of information for the purchaser isn't known until she has the information - but then she has, in effect, acquired the knowledge itself without cost.

Because informational products and processes are constitutive of individuals, communities, and societies, they are constantly interacting with the social, cultural, political, economic, and ecological environments in which they are occurring, which they shape, and to which they refer. Informational products and processes themselves generate additional information. Thus, information commodities serve as materials or even as agents in production processes, something not true either of most material goods or of the other factors of production.

## Information Economy: Alternative Conceptualizations

The three conceptualizations differ in their responses to these problems and, as a result, in how they define the information society. They differ in the aspect of the economy upon which they focus: the first on products, the second on the domain, and the third on functions. They differ in their assumptions and theoretical bases. Inevitably, then, each approach suggests quite different principles, policies, and implementation practices for decision-makers in the public and private sectors. These approaches developed, loosely, chronologically, but are all extant today. They are not mutually exclusive. The earliest continues to dominate policy-making. It is suggested here that enrichment of the third approach (functions) with insights from the first two and other theoretical and empirical material provides the most valid and useful path forward.

## The Information Economy: Products

The first conceptualization of the information economy focuses on the economic role of information and the technologies that handle it as final and intermediate goods and services -on what the economy produces. It defines the information society as that in which those industries which produce information goods and services are proportionately more important to the economy as a whole than they have been in the past. It assumes that the economy works as it always has, and that the neoclassical economic theoretical base remains adequate. The problems experienced by those using neoclassical analytical tools are denied.

This approach arose in the 1960s as part of the self-awareness of qualitative shifts in the nature of society that marked awareness of the emerging information society. This occurred almost simultaneously in Japan, where Umesao coined the term 'informatization' in 1962 and the government subsequently popularized it, and in the US, where Daniel Bell's 1960s notion of the post-industrial society diffused quickly. Both Bell and Umesao took an evolutionary approach to the emergence of the information economy, claiming it was an inevitable stage in the development of societies, following upon the agricultural and industrial stages. There are variations on the theme - some group all activities related to the creation, processing, flow, and use of information into one economic sector, while others distinguish between information and service sectors, and so forth.

This approach became the basis of policy-making once operationalized by identification of the information industries first (at a gross level) by Machlup and then (in a more fine-grained way) by Porat . Porat's technique, which evolved into the system used by the US Department of Commerce and subsequently in the international arena, used Standard Industrial Classification (SIC) codes to identify those businesses and workers within the information sector. Despite the problems of this approach, which necessarily includes some arbitrary decisions (physicians are in, but people who can repair radios are out) and has no way of grappling with many of the significant economic forms informational goods and services can take, it has continued to be the methodology used to derive the percentages of the economy and the labor force in the information sector.

The inadequacies of this approach are increasingly clear. Thus, the Department of Commerce is working to revamp the SIC code system to reflect economic realities, but remains stumped by the same issues that have confounded economists working in this area for over a century. Meanwhile, corporate leaders involved in the "intellectual capital movement"are struggling on their own to develop accounting systems which will permit them to evaluate intangibles as well as the kinds of assets for which accounting tools have been developed.

Despite growing recognition of the limits of this approach, it is likely to continue to dominate policy-making for some time, because of its chronological priority, because it does not require conceptual shifts, because it permits continued use of existing decision-making tools by working policy-makers, and because it serves the interests of some of the most powerful economic entities in the world. Certainly within its terms its insights are valid and important. It does not, however, fully capture many significant qualitative features of the contemporary economy, and thus is inadequate for fully successful decision-making in this environment. Those who take this approach are clinging to an old paradigm in the face of mounting evidence of difficulties in the application of this paradigm. Motivations for continuing to choose this approach in the late 1990s, over 30 years into a discussion of the multiplying problems involved, are multiple.

### The Information Economy: Domain

The second type of conceptualization of the information economy to emerge focuses on the bounding of the domain of the economy itself rather than on relative quantities of different types of goods and services produced within it. Those who take this position define the information economy as that which has expanded through commodification of forms of information never before commodified, including that which is most personal (such as what is in someone's thoughts, or urine) and that which is most public (such as databases created by the government).

This approach received its first full fleshing in the 1970s, again coming from multiple directions. Political economists in the developed world, such as Schiller, Mosco, and Hamelink, began to examine the consequences of the use of new information technologies. Meanwhile, some of those in the developing world who had been struggling for a New International Economic Order since the early 1960s came to realize that a prior condition for redistribution of global material and economic resources was a redistribution of information resources. This view was articulated in such venues as the

United Nations Educational, Social, and Cultural Organization (UNESCO) as calls for a New World Information Order (NWIO) (Richstad and Anderson, 1981). Political economics as applied in this context developed largely through ideas about the nature of dependency and the research it stimulated.

Generally this approach is based on the assumption that the economy has never worked the way neoclassical economists have claimed it worked (to the significant detriment of large portions of the population). Those attempting to use neoclassical economic tools in the analysis of information creation, processing, flows and use have always encountered problems. Economists are simply being forced to acknowledge their existence by the growth of the relative importance of these problems in the information economy.

The weakness of this approach from the perspective of working decision-makers is that its analysis generally remains couched in the broadest theoretical terms (such as, ensure universal access), ignoring the institutional and other mid-level issues that must be dealt with in order to form recommendations for specific policies and implementation practices (such as what access to allow, how access is to be handled institutionally as well as economically and politically, and how access is to be evaluated). Often, those who take this approach call for such radical change - sometimes transformation to a socialist, or gift economy -that discussion becomes polarized and no viable, consensually acceptable, steps appear.

However, the insight offered by those who take this approach is profound. The use of surveillance technologies appears to be increasing rapidly, while concurrently the population of even the United States has less and less access to the information upon which policy-makers base their decisions, and to decision-making processes themselves. This is so for a number of reasons. Shifts in organizational form alter structures of information flows in critical but often obscure ways. Privatization of many formerly public types of information (such as government databases) is a form of disempowerment through commoditization of information. Innovations also extend the ability to treat information and its flows, processing, and use as commodities. The growth in the number of people who want access to the Internet, for example, has itself become a commodity as the addresses needed to identify each individual user (including those within organizations) become scarce and tradeable goods likely to rise in price. Such practices affect our individual and social lives qualitatively as well as economically. Awareness of this expansion of .the breadth of the economic domain and penetration into our lives should be retained by those struggling to deal with the effects of the transformation to an information economy. Those who take this approach recognize the need for a paradigm change and are committed to one among the many alternatives available. It is not yet inevitable or clear around which alternative there will ultimately be coalescence.

## The Information Economy: Functions

The third approach focuses on how the economy functions, and defines this as an information economy because the market has been replaced by harmonized information flows as the key coordinating mechanism. Thus, today's environment is described as a network economy. Those who take this approach acknowledge that the nature of the economy has qualitatively changed. They are building theory based on adapting and developing existing economic concepts as well as inductive and adductive analysis of large amounts of empirical data. A strength of the work of Antonelli , in particular, is that it is based on decades of empirical work examining the many different types of transnational corporations emerging around the world.

Rather than trying to fit new types of phenomena and processes into economic (and legal) categories that no longer fit, those who take the network economics approach, such as Antonelli , Guerin-Calvert and Wildman , and Grabher , are struggling to see in what ways information creation, processing, flows, and use in this environment can be treated like other types of economic entities, and in what ways they cannot. This is a part of a creativity which treats models and typologies as toolkits to be explored for techniques and concepts which can be identified, adapted, discarded, or replaced as needed for particular settings and issues.

In terms of policy, this work represents the merging of two decision-making sectors previously kept distinct: traditionally, corporate strategy was distinguished from public sector policy. Today, firms, networked firms, and government entities often make decisions together in what are called in the literature 'policy networks'. The impact of this merging of types of organizations on the definition of responsibilities of decision-makers to the various communities with which they are involved is neither empirically clear nor normatively resolved. Examples of this approach include identification of a new unit of analysis, the project, involving multiple interdependent organizations, as more useful than either the industry or the firm for analytical purposes. It is precisely because of the harmonization of information flows among these interdependent firms - called network firms or embedded firms - that the role of the market has diminished. In the network economy, a combination of co-operation, co-ordination, and competition replaces sheer competition as the most successful economic strategy. The capacity to innovate, the most valuable capital of all, arises out of networked relations themselves.

Policy principles that flow from this approach are broad, such as the notion that co-operation and co-ordination are as important as competition, and that long-term relationships are more important than maximization of profit from single transactions. Specific policies and implementation practices are as likely to come from the field of organizational sociology as from public administration. McLuhan noted that the content of each medium is another medium, often the medium which just preceded it in dominance. From this perspective, network economics is about the global information infrastructure as medium and the economy as content. This emphasis on information flows does not deny the continued significance of capital as a motive force and as an operating framework - but capital itself is heterogeneous, and new forms are emerging.

In terms of a Kuhnian paradigm change, this approach to the conceptualization of the information economy offers the greatest validity and utility, and thus is most likely to provide •the seed from which a new paradigm for understanding the economic world will grow. It is also interesting from the perspective of the sociology of knowledge because it highlights the importance of distinguishing among different manifestations of a paradigm and their relative importance. While Antonelli quacks like a mainstream economist quacks, and walks like a mainstream economist walks, the content of his insights is in many ways more genuinely radical than those of Marxism-driven proponents of the second approach.

## Network Economics Approach

Network economics might usefully gain from incorporation of insights offered by the other approaches to conceptualizing the information economy. The first perspective, focused on information products, is forced upon the network economics world through government policy based upon it and related accounting schemes. In turn, however, the network economics approach is likely to be of greater and greater influence during struggles like those currently going on over reformulation of the SIC code and related efforts in the accounting world.

It may be more difficult to incorporate into network economics the insights of the second approach into the nature and effects of expansion of the economic domain. Doing so requires finding ways of either adapting or developing new decision-making processes so that they can take qualitative types of information into account, or find ways of measuring things generally considered to be non-quantifiable. One example of the latter can be found in the efforts over the past half dozen years to develop ways of applying intellectual property rights to traditional bodies of knowledge held by indigenous tribal groups, and to knowledge held by communities; another is the development of tools to quantify 'wildlife rights' so that they can be incorporated into the economic system.

The network economics approach can also be usefully enriched by incorporation of the work of Scazzieri, who offers an original conceptualization of the nature of production processes which provides a superb fit with current conditions. As Scazzieri notes, under current conditions, analyses must deal with those production factors, processes, products, and markets that may be virtual as well as those that are potential and actual. He also distinguishes among types of production processes in a way that is particularly useful in the digital, network environment. Elementary processes are elements out of which complex processes are comprised; and scale is the number of elementary processes involved in a particular production process. In the information economy, scale can be understood in terms of an information production chain which includes information creation, processing, storage, transportation, distribution, destruction, seeking, and use. Actors in the global information economy, then, are currently manipulating electronic, legal, political, cultural, and organizational infrastructures for digital activities as they seek to expand their scale.

Further elaboration of what an enriched network economic approach might look like must involve cross-checking what is incorporated into network economics with those various literatures that touch upon information creation, processing, flows, and use in the history of economic thought. The result would outline a research agenda for those working in this emergent area of economics, whether it is ultimately known as network economics or as the economics of information. By either name, this branch of economics will supply the foundation upon which policies will be made and strategies developed in the information economy. It is hoped that the conceptual clarifications offered here will, in turn, participate in the co-evolution of conceptualizations of the information economy and in the shaping and growth of the field.

## References

Antonelli, C, (1992), The economic theory of information networks. In: C. Antonelli (ed.), *The Economics of Information Networks*. North-Holland, Amsterdam., 5-27.

Babe, R. E., (1995), *Communication and the Transformation of Economics: Essays in Information, Public Policy and Political Economy*. Westview, Boulder, CO.

Hopwood, A. G. & P. Miller (eds), (1995), *Accounting as Social and Institutional Practice*. Cambridge University Press, Cambridge.

Lamberton, D. M. (ed.), (1974), *The Information Revolution. Annals of the American Academy of Political and Social Science,* 412.

Stiglitz, J., (1985), Information and economic analysis: a perspective. *Economic Journal,* 95, 21-41.

Temin, P. (ed.), (1991), *Inside the Business Enterprise: Historical Perspectives on the Use of Information.* University of Chicago Press, Chicago.

# 7

# Impact of ICT on Production Efficiency

The information and communication technologies (ICTs) have evolved, improved and become more widespread and constitute a very dynamic sector of economic activity. Developing countries' exports of ICT goods and services have grown spectacularly, establishing new international specialization patterns. At this stage it is important to establish to what extent ICTs have impacted on economic growth. The analysis of differences in economic gains from ICT between countries, industries and firms aims at revealing where most positive effects have occured and where new efforts should focus.

The focus on productivity and the supply side is of interest to developing countries. In recent years, developing countries have continued to grow extensively by absorbing new investment and labour. In particular, the policy choice to build up and support an ICT-producing sector has generated jobs and attracted additional foreign investment and offshoring in several emerging economies. In addition to recognizing the benefits of the ICT-producing sector, it is important to recognize the development potential of other sectors of economic activity where ICTs could be used intensively.

An important result of the intellectual debate initiated by Gordon was that in developed countries, while initial productivity gains from ICT accrued mainly from a fast growing ICT-producing sector, in the second stage there were additional gains deriving from increasingly widespread use of ICTs in many economic sectors. In developing countries, by contrast, productivity gains from ICT are still being largely generated by the ICT-producing industry This is because developing countries with tighter budget constraints have been spending less on ICT. In that context, developing countries wishing to increase their gains from ICT should take measures to extend the use of ICT from the ICT-producing sector to other economic sectors and to domestic consumers.

Researchers studying the impact of ICTs on productivity used different economic models of interpretation, depending on the type of ICT data made available and the issue researched. The analysis depended very much on the quality of measurements and the availability of data on ICT uptake. At the firm level, the specialized literature tended to move away from evaluating the impact of all-inclusive ICT investment and instead focused on the way in which specific ICT use resulted in higher production efficiency. As ICTs become more complex and widespread, more information is needed on how ICT use determines competitiveness, in addition to information on the amount spent on ICT investment. In developing countries as well, more information on how firms use ICT would help in refining policy decisions to scale up the most efficient enterprise practices.

## ICT Capital, ICT Use and Growth

Measuring the amount of capital invested in ICT goods and services is one way of measuring the intensity of ICT uptake in an economy. Within the neoclassical growth accounting setting, ICT investment is assumed to contribute to labour productivity through capital deepening, this assumption being consistent with the logic that workers who have more computing and communication equipment are more efficient than workers who have only ordinary capital. Thus ICT capital contributes to GDP per capita growth as an additional input to the production function. This theoretical framework does not account for the contribution of ICT to total factor productivity arising from an ICT-enabled innovative reallocation and reorganization of production (such as the substitution of ICT capital for ordinary capital). Several such growth accounting exercises were conducted by a number of authors over different periods of time to estimate the contribution of ICT capital to GDP growth.

Gains to productivity from ICT have low values when ICTs are nascent in the economy. When ICT inputs have a very small share in total production, growth accounting models can detect only a feeble effect on productivity from those inputs. For example, early results in developed countries derived for the 1980s and early 1990s showed that investment in computer equipment had a minimal impact on GDP growth. Research estimating the impact of investment in computing equipment on GDP growth in the United States in the 1980s found a rather small contribution of just 0.2 per cent compared with an average GDP growth rate of 2.3 per cent.

The ICT variable used in that research was the yearly variation in the stock of computer capital. The broader concept of ICT investment was developed later, in parallel with research and has been used only recently. It was suggested by several authors that inputs with a very small share in total output could not have a great impact on GDP even if investment in that input was growing at high rates.

Although computer investment had reached a sizeable amount in certain specialized sectors, at the national level investment in computing equipment represented only 0.5 per cent of the total capital stock in 1993. This measurement problem was subsequently addressed by broadening the concept of computer capital to also incorporate other ICT goods and services, including software and communication equipment. Other measurement problems related to accurately capturing price changes in IT and composite IT goods and services since the rather insubstantial changes in nominal ICT investment were hiding much larger variations in real ICT investment.

Conversely, when specific ICTs such as computers become ubiquitous, especially in developed countries, they can no longer contribute to explaining variations in output between economic actors. In that case, it is necessary to find other measurements relating to a particular use of ICT (for example, establishing web presence) where significant differences in ICT use explain changes in output.

An additional insight into growth accounting for ICT was provided by the structural change in the mid-1990s and the shift in the trend of ICT investment. Starting in 1995, there was a marked decline in global ICT prices and a considerable acceleration of investment in ICT. At first, this trend was apparent only in developed countries. With a time lag of almost a decade, developing countries also experienced a similar increase in their ICT capital investments. In the mid-1990s developing countries started to invest in ICT at faster rates and some, notably from Asia, have acquired a considerable stock of ICT capital that matches ICT capital intensity in the United States in the early 1990s. In that connection, the analysis of more recent data on ICT investment in developing countries benefits from comparisons with ICT investment trends in developed countries in the 1980s and 1990s.

Using the new data from after 1995, Jorgenson and Vu found that the contribution of ICT capital to world GDP had more than doubled and now accounts for 0.53 per cent of the world average GDP growth of 3.45 per cent. The percentage was higher for the group of G7 countries, where ICT investments contributed with 0.69 per cent to a GDP growth of 2.56 per cent during 1995–2003. The Economic Commission for Latin America and the Caribbean has recently conducted a series of growth accounting exercises for Latin American countries. Their estimates are reflected below for a comparison with certain Asian developing countries.

Economies appear in increasing order of the contribution of ICT capital to growth, ranging from 0.09 per cent in Indonesia to 0.69 per cent in the G7 countries and reaching 0.38 per cent in G7 countries prior to 1995. In a number of developing countries, accelerated investment in ICT made a significant and positive contribution to GDP growth after the mid-1990s. However, owing to the low starting levels of ICT capital, this type of input could make only a reduced contribution, below 0.25 per cent in many cases. It was necessary for ICT capital stock to reach a certain level before greater benefits from investment in those technologies could be achieved. There were exceptions to that general finding, however. In the Republic of Korea, Costa Rica, Chile and China benefits from ICT investment were about 0.5 per cent, closer to the G7 average. In three out of those four developing countries, the ICT-producing sector had a large share in economic output.

Several emerging economies have experienced strong growth in the last 10 years. In fast-growing developing countries ICT investment made only a small contribution to GDP growth. Other inputs such as ordinary capital investment and labour remained major contributors to the rapid GDP growth in emerging economies. China and India stood out with some of the largest output growth in recent times, but their performance was led only partly by greater investment in ICT, as the other inputs distinguished by the model had a considerably higher share. Previous editions of the ECDR have shown that the effect of ICT use on other fundamental variables of the economy such as labour was also significant. It is therefore possible that in emerging economies ICT use made a larger contribution to economic growth than suggested by the capital-deepening effect. Indeed, ICT use has also contributed to improving labour productivity in emerging economies. However, as growth accounting estimates show, developing countries in which ICT investments had a greater impact on growth also managed to leverage more capital and to improve the skill mix of their labour force. Benefits from ICT capital did not occur in isolation from a general upturn in the economic evolution of emerging economies.

## The ICT-producing Sector and the ICT Using Sectors

In a debate initiated by Gordon, researchers asked whether there was proof of gains from the widespread use of ICTs in addition to gains to GDP from rapid technological advances in the ICT-producing sector. For the United States economy, Stiroh showed that gains from ICT investment in a broad range of economic sectors were positive and complementary to gains from the ICT-producing industry itself. Furthermore, in a comprehensive empirical study at the sectoral level comparing productivity growth in the United States with that in the European Union, van Ark, Inklaar and McGuckin found that the faster productivity growth in the United States is explained by the larger employment share in the ICT-producing sector and productivity growth in services industries with a highly ICT-intensive profile. Accordingly, wholesale and retail trade and the financial securities industry account for most of the difference in aggregate productivity growth between the EU and the United States.

This debate is of relevance to development because in many developing countries ICT investment has remained largely concentrated in the ICT-producing sector. Several studies have shown that while in developed countries, the impact of ICT use was an important element accounting for growth, in most developing countries productivity gains were generated predominantly by the ICT-producing industry itself. On a global scale, developing countries managed to attract investment and started to produce a large share of global ICT goods and services. However, demand for ICTs originated predominantly from developed countries. This pattern has recently changed in the case of certain developing countries. For example, after a rapid development, China became in 2006 the world's second largest importer of ICT goods, with a large share of those imports originating from developing economies of Asia.

Bayoumi and Haacker compared gains to real GDP from the IT-producing sector with gains to total consumption from domestic IT spending

(1996– 2000). They found that several developing economies received a smaller demand boost from IT spending even if they were relatively specialized in IT production. Because ICT use in several developing economies is limited to a few economic sectors those economies gained less from ICT. IT spending had a proportionately greater impact on demand in Ireland and Finland as compared with the effect in the Philippines and Thailand, for example. Among developing countries specialized in the production of ICTs, the Republic of Korea, Singapore and to a certain extent also Malaysia had a larger contribution of IT spending to domestic demand growth. However, as the authors suggest, developing countries gained from IT mostly as IT producers, while developed countries gained relatively more as IT consumers.

There are several factors that can explain why firms in developing countries use ICT less. Data on ICT use by firms in Thailand, for example, show that the main factors hindering the adoption of ICT applications are high costs and lack of knowledge about how to apply ICTs for improving performance in a particular type of business. Undoubtedly, the lack of knowledge about how to apply ICT to business goes hand in hand with the reduced size of e-markets in specific sectors. Certain domestic producers continue to be confronted with lower levels of ICT penetration among domestic suppliers and consumers. Owing to lack of knowledge about how to implement ICT in order to boost sales, domestic producers find it best to refrain from adopting ICTs.

Furthermore, data on Thai firms show that exporting firms connect their computers to the Internet and establish web presence more often than firms producing for domestic markets, so as to access information and reach consumers in foreign markets. This is confirmed by research findings in Clarke and Wallsten, which show that the Internet affects exports from developing countries in an asymmetric manner. Their study measures ICT adoption by the number of Internet users within a country (the Internet penetration rate). They find that developing countries with higher Internet penetration rates tend to export more merchandise to partners from developed countries, while there was no significant relationship between Internet access and trade in goods between developing countries. This finding indicates that in developing countries the Internet has been used more effectively in trade relations with developed countries.

## ICT Impact on Total Factor Productivity Growth

Atkinson and McKay describe three ways in which ICT can contribute to total factor productivity: network externalities, complementary improvements generated by ICT adoption, and improved access to knowledge. All three suggest that the positive effect of ICTs on productivity will occur with a time lag. ICTs can generate network externalities in the same way as connecting a popular service provider to the telephone network will increase the satisfaction of all telephone subscribers. However, it goes without saying that building valuable networks of ICT users takes time and the process is not without difficulties. For example, the different technologies involved are not always compatible and interconnections may fail. The same applies to organizational change – which is costly, time-consuming and sometimes unsuccessful – and also to accessing information: not all ICTs are equally user-friendly and users may find it hard to identify the best source of information available.

Estimates of the effect of ICT on total factor productivity growth are provided either by growth accounting models or by other regression models explaining total factor productivity growth. Total factor productivity growth reflects technical change in the economy that cannot be explained by mere increases in capital and labour, but it also captures other factors such as policy and institutions. There are several ways in which ICTs could contribute to total factor productivity growth, but so far there is no agreement on the size of this contribution. More rapid technological progress in the ICT-producing sector is one way in which ICTs influence total factor productivity. For example, Oliner and Sichel ran a sectoral growth accounting model on data corresponding to the non-farm business sector in the United States and found that both the contribution of IT use to labour productivity and the total factor productivity growth arising from the IT-producing sector are positive and statistically significant.

IT capital investment accounted for the larger share of labour productivity growth in the United States through a general IT capital deepening effect in several economic sectors. The IT-producing sector had a relatively lower contribution to labour productivity in the United States, which was above all influenced by productivity gains in the semiconductor industry. Other growth accounting exercises were performed with European data and with East Asian data.The ICT-producing sector had a sizeable positive effect on total factor productivity growth in several East Asian economies specialized in the production of ICT, and also in Ireland. In those economies, productivity growth in the ICT sector was the main source of technological improvement and gains to GDP from ICT production exceeded gains from ICT use (capital deepening).

Research has also investigated whether ICT-using industries have contributed to total factor productivity growth. Schreyer, for instance, developed a growth accounting model in which ICT capital contributed to GDP growth through both capital deepening and spillover effects as a source of technological improvement. However, to date there is no agreement about whether ICT investment has a significant effect on total factor productivity growth in developing countries. More research is needed on this topic.

There is an emerging literature attempting to estimate the contribution of ICT to total factor productivity growth in developing countries. For example, Shiu and Heshmati estimated the impact of foreign direct investment and ICT investment on total factor productivity growth in China over the last decade (1993–2003). While Chinese total factor productivity grew at an average rate of 8.86 per cent to 9.22 per cent during that decade, ICT investment was found to contribute to it to a lesser extent: a percentage increase in ICT investment resulted in 0.46 per cent total factor productivity growth, compared with foreign direct investment that generated 0.98 per cent.

Seo and Lee estimated a model of total factor productivity growth during 1992–1996 in 23 OECD and 15 non-OECD developing economies as a function of GDP growth, human capital, openness, domestic ICT capital intensity, ICT spillover effects and a linear trend. In their model, ICTs can have a double effect on total factor productivity growth – through higher domestic investments and through network effects. The ICT network variable is calculated as the ICT capital intensity of foreign countries. Their analysis compared two samples of OECD economies and developing economies. Findings suggest that in developed economies total factor productivity growth is more strongly correlated with domestic ICT investments, while in developing countries it is not significantly linked to domestic investments in ICT but rather to ICT investments in OECD countries.

The results reviewed confirm that gains from ICT to GDP occurred both in developed and in developing countries, with a lower intensity in countries that started to invest in ICT later and at a slower pace. The positive macroeconomic impact of ICT coincided with a marked contribution to growth deriving from other factors such as increased ordinary capital investment, improved skills and a better allocation of resources. A number of developing countries, particularly Asian ones, have specialized in the production and exporting of ICTs. As a result, it is estimated that ICT production increased GDP by as much as 7 per cent in Singapore and 3 per cent in Malaysia yearly (1996–2000). Developed countries gained more than developing countries from the consumption of ICTs. In terms of impact of ICT investment on total factor productivity, as a sustainable factor for growth, research results do not always confirm the existence of significant effects. However, when a positive effect is verified, it is explained by a more efficient use of ICT in specific sectors (the case of wholesale and retail trade services in the United States). It is therefore necessary to study in detail how the specific use of ICT leads to greater business efficiency.

## Firm-level Impact of ICTs

This section draws on a growing number of empirical studies at the firm level conducted in a number of OECD countries. In the context of the Partnership on Measuring ICT for Development, several international institutions have pledged to work on improving the information available on ICT measurement and the quality of the data, and to make ICT statistics cross-country-comparable. As the next step, the OECD and Eurostat are promoting research to estimate the economic impact of ICTs at the macroeconomic, industry and firm levels. Initiated in 2003, the EU KLEMS project aims at building a comprehensive time series database, including measures of economic growth and productivity derived through the use of growth accounting techniques.

Through the Partnership on Measuring ICT for Development, UNCTAD has worked to improve the measurement of ICT use by businesses in developing countries. In those countries there is little research estimating the impact of ICT on production efficiency at the firm level. In order to conduct that type of study, statistical offices or research institutes need to collect data on the use of ICTs and link this information with output and productivity data at the firm level. While several developing countries have made important progress in measuring and producing data on ICT use, linking this information to other economic variables often has remained a challenge.

Investing in ICTs makes a difference to firms' sales and labour productivity. Although ICT investment cannot confer long-term competitive advantage, not investing in ICT clearly leaves firms at a disadvantage.

Several recent studies highlight the need to complement data on ICT investment with more information on the way in which firms use specific ICT. Particularly in a developed country context, investment in ICT has become nearly universal and spending more on acquiring additional and more complex technology may not lead to increased market shares in firms. Accordingly, it became increasingly relevant to study the way in which ICT use by firms is related to higher sales per employee. In developing countries, the percentage of companies with access to ICTs remained lower on a national scale, even though for certain economic sectors or geographical regions indicators showed higher ICT penetration.

### ICT Use and Firm Labour Productivity

The different dimensions of firm-level analyses considerably increase the complexity of research results on this topic. If previously presented models focused mostly on ICT investments, the studies reviewed here go into greater detail about specific ICT technologies or applications and interactions between them. Moreover, variables on ICT use have different productivity effects in conjunction with other control measures such as firm age, the share of foreign capital participation or access to more skilled human resources. To make it easier to keep track of the different empirical model variants, this section groups the many dimensions into three categories, by elements of the regression equation.

Labour productivity is commonly measured either as value added per employee or as sales per employee. Criscuolo and Waldron derive results for both measures of productivity and find that the impact of e-commerce was slightly stronger on value added than on sales. Conversely, Atrostic and Nguyen rely on the findings of Baily to argue that using value added as a measure of labour productivity yields systematically biased estimates of the theoretically correct growth model. Outside the context of empirical models, value added is a more precise measure of labour productivity since it subtracts from the value of sales the costs incurred with intermediate consumption. However, as suggested by Carr, firms take decisions to adopt ICTs on the basis of expectations about maintaining or increasing market share and competitiveness. For those considerations, analysing about the impact of ICT on sales per employee is more appropriate.

The empirical models presented here draw on several types of variables for describing the use of ICT. Binary variables, which distinguish between firms with and without access to, for example, Internet, are easy to collect and provide input for analyses of differences between the haves and the have nots. Empirical studies also test for the effect of intensity in ICT uptake – for example, the share of capital devoted to computer investments – and intensity of ICT use, such as the number of computers available or the share of employees using e-mail. From a theoretical point of view, findings based on numerical rather than binary variables are more powerful. Maliranta and Rouvinen proposed a slightly different modelling structure in which they estimate the net effects of several complementary features of computers: processing and storage capacity, portability and wireless and wireline connectivity. In their model, the positive labour productivity effect associated to the portability of computers is complementary to that of the basic processing and storage capacity of any computers whether portable or not.

The different variables of intensity in ICT use by enterprises analysed by the specialized literature are not the ideal measurements. As some have emphasized, ICT use becomes increasingly relevant to productivity when combined with soft skills such as good management and superior marketing abilities. Unfortunately, such soft skills and soft technology inputs cannot be quantified directly and therefore their effect is hard to assess. Empirical research usually corrects for this unknown effect by accounting for different economic results in foreign-owned firms, in exporting companies, in establishments belonging to multi-unit corporations or simply in more experienced firms. Therefore, policy implications derived from such research do not directly recommend that the intensity of ICT use be scaled up (for example by increasing the number of computers per employee). Rather they recommend investigating how the combined use of ICT and superior managerial capabilities can account for variations in ICT gains between firms with different characteristics.

ICTs can generate higher market shares either by reducing input costs and thus allowing firms to produce more of the same products, or by improving the quality of products or product packages, with, as a result, additional sales or higher-priced products. Empirical results presented here cannot distinguish between those two effects. More information on the evolution of prices in different sectors is needed in order to assess which effect prevailed in defined periods of time.

**Complementary Factors Explaining the ICT-productivity Relationship**

Control variables are additional elements in the growth equation. They also give a different dimension to results relating to ICT use and productivity when interacted with measures of ICT.

In several studies, firm age has proved to be an important element explaining productivity effects. The European Commission's Enterprise and Industry Directorate General showed in a report that the dynamic evolution of new firms is a source of economic growth and employment and that new firms also contribute significantly to the diffusion of e-business applications in Europe. In terms of econometric results, Maliranta and Rouvinen estimate that young manufacturing firms in Finland, unlike older ones, have 3 per cent higher productivity gains from the use of computers. Also, young Finnish services firms appeared to be 1 per cent more productive thanks to access to the Internet. In a different study, Farooqui runs four different growth models on young and older British firms in manufacturing and services taken separately. Results show that ICT indicators such as investment in IT hardware and software and the share of ICT-equipped employment have a more pronounced impact on young manufacturing firms as compared with older ones. The same finding did not apply to young British services companies, however, on that issue, Atrostic and Nguyen draw attention to the fact that the measure of capital input used in most papers – the book value of capital – is a more accurate proxy in the case of the new firms. Older firms' capital input is not properly captured by book values because this measure is evaluated at initial prices when capital assets were acquired as opposed to current asset prices. The first best proxy to use would be the current value of the capital stock computed by means of the perpetual inventory method by using information on yearly capital investments, depreciation and current asset prices. But in many cases, data are not available on all the above-mentioned variables. Regression results using the book value of capital assets are likely to give biased results for older firms and more accurate results for younger ones.

Firms with foreign capital participation seemed to have higher labour productivity. With regard to developed countries, Bloom, Sadun and van Reenen estimated that in their large sample of UK firms from all business sectors, US-owned establishments had significantly higher productivity gains from IT capital than other foreign-owned firms or domestically owned firms. This result can be linked with macro-level findings which indicated that the

United States had acquired greater labour productivity from investment in ICT than all other developed countries, especially since the mid-1990s. More productive US-owned firms appear to be better managed or have access to more efficient ICT solutions.

Similarly, firms belonging to multi-unit networks of affiliates may have greater labour productivity since they dispose of additional resources to draw from in the subsidiary–headquarters management structure. A multi-unit corporate configuration may justify benefits from network effects (a success story replicated in several subsidiary branches) and access to superior management resources.

The skill mix of production and non-production workers and the level of education in the regions where companies are located were considered by some studies to be complementary to the measures of ICT use by firms. Better-skilled workers are more likely to be able to develop, use and maintain more advanced technology. Maliranta and Rouvinen comment that growth models need to control for the human capital characteristics of employment and labour because these variables are essentially complementary to ICT uptake and omitting them would innate the labour productivity gains from ICT.

Last but not least, when quantifying the relationship between ICTs and labour productivity one needs to control for differences in demand and supply factors. For example, in many countries businesses located in the vicinity of the capital benefit from higher demand than those located in isolated provinces simply because there is a high concentration of the population in capitals. In a similar way, different industries have distinct labour productivity averages owing to both demand and supply factors. For example, an oil-producing company is very likely to have higher sales per employee than a light industry manufacturer specialized in food and beverages of the same size (because of industry characteristics such as price, labour intensity and type of consumer good). It is therefore necessary to take into account regional and industry-specific characteristics when accounting for the contribution of ICT to labour productivity growth.

The ICT-producing sector itself benefited from ICT use that considerably exceeded domestic industry averages. Maliranta and Rouvinen estimate that in Finland firms belonging to the ICT-producing sector had 3 to 4.5 per cent higher labour productivity gains from ICT use than the rest of the manufacturing and services companies in the sample. This may be because ICT producers have a know-how advantage over other ordinary users in terms of how to best put to work specific technology to enhance labour productivity.

## Impact of Specific ICTs on Productivity

Use of computer networks (such as the Internet, intranet, LAN, EDI and Extranet) had an estimated 5 per cent positive effect on labour productivity in a large sample of American manufacturing businesses. The model considered a theoretical framework in which use of computer networks made a "disembodied" contribution to technological change other than that of capital and labour. Atrostic and Nguyen take up again the impact of computer networks in a slightly modified empirical model. The novelty of their approach consists in using two different computer-related measures in the labour productivity regression: computer capital, as distinct from ordinary capital, and the computer network binary variable used previously. In their view, having separate measures for the presence of computers (computer investment) and for how computers are used (computer networks) is crucial for estimating accurately the two effects on labour productivity. When using a sample composed only of newly registered US manufacturing firms, they find that the contribution of computer networks added 5 per cent to labour productivity while investment in computers added 12 per cent. Within the entire data set of older and younger US manufacturing firms, the contribution of computer capital dropped to 5 per cent and there was no evidence of a positive effect on computer networks any more. However, as mentioned before, most empirical studies tend to find that ICT use has less impact on older manufacturing firms, and this may be due to a measurement bias as explained in Atrostic and Nguyen.

E-commerce also has a significant impact on labour productivity in firms, with a marked difference between businesses that buy and those that sell online.

Criscuolo and Waldron analyse a panel of UK manufacturing firms and find that the positive effect of placing orders online ranged between 7 and 9 per cent. On the other hand, they estimate that firm labour productivity was lower 5 per cent for those that used e-commerce for receiving orders online (online sellers). Lower labour productivity associated with selling products online is likely to be due to price effects. The prices of products sold online are considerably lower than the prices of similar goods sold through different channels. Additionally, firms which specialize in selling online may have difficulties in finding suppliers from which to buy online as much as they would want. Larger firms with a stronger position in the market may be able to better cope with balancing the extent of e-buying and e-selling In a larger and updated UK data set of manufacturing and services firms, Farooqui finds again that e-selling negatively impacts on labour productivity in manufacturing, while e-buying has a larger and positive effect. In particular, Farooqui finds that in distribution services e-buying boosts labour productivity by 4 per cent.

Several studies show that measures of ICT use by employees are also reflected in enhanced firm productivity. Within a large panel of Finnish firms Maliranta and Rouvinen compare the impact of computer use on labour productivity in manufacturing and services sectors. A 10 per cent increase in the share of computer-equipped labour raises productivity by 1.8 per cent in manufacturing and 2.8 per cent in services. On the other hand, a higher share of employees with Internet access was found to have a significant impact only on services firms (2.9 per cent). The study considers Internet use as a proxy for external electronic communication and LAN use as a measure of internal electronic communication. Findings show that manufacturing firms benefit more from better internal communication, while services firms gain more from improved external electronic communication. A 10 per cent higher share of employees using LAN in the manufacturing sector results in 2.1 per cent higher labour productivity. A similar study on a mixed sample of Swedish manufacturing and services firms estimated that a 10 per cent higher share of computer-equipped labour boosts productivity by 1.3 per cent.

The Swedish study also estimates productivity effects deriving from access to broadband of 3.6 per cent. With the help of a composite ICT indicator, Hagén and Zeed show that adopting an ever-increasing number of ICT solutions has positive but decreasing effects on labour productivity in Swedish firms. Each additional level of ICT complexity seems to add less to firm productivity. Farooqui also identifies the use of computers and the Internet by employees as a proxy of work organization and skills. In the United Kingdom, a 10 per cent increase in the share of employees using computers raised productivity by 2.1 per cent in manufacturing and 1.5 per cent in services. These effects are additional to the impact of ICT investment, also accounted for in the Farooqui model. Similar estimates for Internet use by British firm employees showed 2.9 per cent for manufacturing and no significant impact for services.

Maliranta and Rouvinen estimate the impact of different complementary computer features on labour productivity in a 2001 sample of Finnish services and manufacturing firms. Their computer variables are measured in terms of share of employees using computers with one or several of the following features: processing and storage capabilities, portability and wireline or wireless connectivity. They find that a 10 per cent higher share of labour with access to basic computer attributes such as providing processing and storage capabilities increases labour productivity by 0.9 per cent. In addition, computer portability boosts output per employee by 3.2 per cent, wireline connection to the Internet adds 1.4 per cent, while wireless connectivity adds only 0.6 per cent.

There is an emerging literature estimating the impact of broadband on firm productivity. Gillett et al. were the first to quantify the economic impact of broadband and found that there were positive and significant effects on the number of workers and the number of businesses in IT-intensive sectors.

**ICT Investment, Soft Technologies and Total Factor Productivity Gains**

Brynjolfsson and Hitt explore the impact of computerization on multi-factor productivity and output growth in a panel data set of 527 large US firms over a period of eight years (1987-1994). They compare the short-run and long-run effects of computerization on total factor productivity growth by taking first and five- to seven-year differences of the log-linear output function. The aim of their analysis is to understand the mechanism through which private returns from computerization accrue, since they are the ultimate long-run determinants of decisions to invest in ICTs.

They model production as a function of computer capital and other inputs, and assume that in the presence of computers, the efficiency of employees, internal firm organization and supply-chain management systems are improved. They find that, in the short run, investments in computers generated an increase in labour productivity primarily through capital deepening, and found little evidence of an impact on total factor productivity growth. However, when the analysis is based on longer time differences, results show that computers have a positive effect on total factor productivity growth. In the long run, the contribution of computer capital to growth rises substantially above computer capital costs, and this is then reflected in terms of total factor productivity. Their interpretation is that computers create new opportunities for firms to combine input factors through business reorganization. Brynjolfsson and Yang had previously estimated that computer adoption triggers complementary investments in "organizational capital" up to 10 times as large as direct investments in computers.

More research, based on developing country data, is needed in order to ascertain how and when ICT use increased production efficiency in firms and which ICT was used. A comparison of estimation results across different countries, industry sectors and technologies can provide policymakers with additional information for fine-tuning ICT policy master plans.

## References

Atkinson RD and McKay A (2007). *Digital Prosperity: Understanding the Economic Benefits of the Information Technology Revolution.* Washington, DC: Information Technology and Innovation Foundation.

Baily MN (1986). Productivity growth and materials use in US manufacturing. *Quarterly Journal of Economics,* Vol. 101, No. 1, pp. 185–196.

Carr N (2004). *Does IT Matter? Information Technology and the Corrosion of Competitive Advantage.* Harvard Business School Press.

Criscuolo C and Waldron K (2003). *E-Commerce and Productivity.* Economic Trends 600, UK Office for National Statistics.

UNCTAD (2008). *The Impact of ICT Use in Manufacturing Firms in Thailand.* Joint UNCTAD-Thailand National Statistical Office project.

8

# E-Business and Innovation Policies

As economies, mostly in the developed world, but increasingly in developing countries too, move towards an even more knowledge-intensive stage of evolution, understanding the role of the technologies that facilitate the accumulation, diffusion and absorption of knowledge in innovation-driven competition becomes crucially important. In particular, knowledge has long been recognized as one of the key factors underpinning innovation and, through it, economic growth and development in market economies.

## Importance of Innovation Policies

Innovation, in a narrow sense, can be defined as the introduction by an enterprise of an entirely new product, service or productive process not previously used by anybody else. In a broader sense, the concept of innovation can be applied to the activities of enterprises which introduce for the first time products or processes that were not present in the particular market in which they operate, or in certain contexts, were merely new to the firm. In that broader sense, innovation can also take place when new products or processes spread through imitation or when gradual improvements resulting in higher productivity are made. Innovation matters because, thanks to its positive effects on productivity, it is the major driver of long-term increases in output per worker and hence it determines the ability of the economy to support better living standards for all.

Public policies in support of the innovative activities of enterprises are justified by the existence of a number of possible market failures that could result in private investment in innovation that would fall below the socially desirable levels. Such market failures can affect both financial markets and markets for goods and services. Intellectual property regimes and cross-border investment rules also play fundamental roles in the generation and diffusion of knowledge and innovation. The existence of a strong national scientific base is another crucial reason for policy interventions to support innovation.

R&D is often highly capital-intensive. At the same time, the uncertainty of innovation and the existence of large asymmetries in the information that is available to innovators and to the providers of capital make it more costly for innovative enterprises, particularly the smaller ones, to access capital to fund R&D activities and/or the practical implementation of their outcome. The fact that knowledge spillovers make it impossible for the financial returns of innovations to be fully appropriated by innovative firms adds to the risk that the operation of financial markets may result in suboptimal levels of R&D investment.

Failures that affect the operation of markets for goods and services may also hold back innovation. Competitive pressure is a strong motive for innovative behaviour among enterprises; removing factors that unnecessarily restrict competition may therefore provide a way to accelerate innovation. The reduction of the impediments that limit market size and hence the opportunities for innovators to recoup the sunk cost of R&D activities is another measure to consider in innovation-supporting policies.

Patents, the main instrument of intellectual property regimes, offer innovators a time-bound monopoly in the commercial application of the knowledge they produce. This enables them to recover the investment that R&D activities represent and may thus provide an incentive for innovation. However, innovation can happen, and has happened, in the absence of strong patent protection. A strong patent regime makes it more costly for other potential innovators to build on existing knowledge in order to generate their own innovation. The balance between the benefits and costs of patents from the point of view of the public interest (i.e. to support innovation and to facilitate knowledge spillovers) is not a straightforward one.

Knowledge spillovers may derive from the activities of publicly funded academic research centres and also from private-sector, commercially motivated research. Transnational corporations tend to be more active and obtain better results in their innovation efforts than domestic firms. But it does not necessarily follow from this that policies to attract foreign direct investment (FDI) are the only, or even the best, instruments that developing countries have to facilitate the dissemination of innovation spillovers in their economies, because (a) transnational corporations are also better than domestic firms in preventing knowledge spillovers, and (b) domestic enterprises in developing countries often lack the capacity to absorb potential spillovers without outside support. In order to diffuse knowledge and innovation across the economy, FDI policies are therefore often complemented with interventions to facilitate the interaction between local firms and innovative international firms. From the point of view of the reinforcement of the absorptive capacity of local firms, it is important to facilitate the emergence of a solid scientific base at the national level. Even in a highly globalized world, knowledge transmission often takes place within relatively small networks and does not always flow easily across borders.

The application of ICTs to productive, financial, administrative and commercial activities has been particularly fertile in terms of innovation. ICT use by enterprises has helped them to become more efficient through business-process innovations (for example, electronic data interchange), and it has also resulted in the emergence of entirely new products or services (for example, online banking).

The importance of the dynamic relationship between the use of ICT and innovation is increasingly recognized at the policy level. A sign of this is the abundance of initiatives at all governmental levels aimed at supporting ICT-driven innovation. Another one is the emerging trend for policies to promote ICT use by enterprises on the one hand and policies to foster innovation on the other hand to be entrusted to the same policymakers or placed under the same overall political responsibility. For example, in around half of the countries covered in a recent survey by the European Union the ministry or agency that is responsible for innovation policy is also involved in e-business promotion. Most often, innovation policies include or address aspects of the use of ICT by businesses. However, even when innovation and policies related to the use of ICT by enterprises share the same institutional framework, it is not necessarily the case that they are envisioned as a single set of policy objectives with a coherent arsenal of policy instruments to achieve them. The borderlines are uncertain and ministries and agencies dealing with matters such as industry, small and medium-sized enterprises (SMEs), education, scientific research and others may be involved at various levels.

## Impact on Innovation of the Use of ICTs by Enterprises

Economic globalization has significantly increased the competitive pressure on enterprises in many sectors. This comes as a result of, among other factors, the emergence of new, lower-cost producers, fast-changing demand patterns, increased market fragmentation and shortened product life cycles. In such an environment, innovation (either in terms of business processes or of final products and services) becomes crucial for the long-term competitiveness and survival of enterprises. It also enables them to climb the value ladder, a particularly important consideration for the enterprises of many developing countries. At the same time, the enterprises of those countries, particularly SMEs, face serious difficulties in benefiting from ICT-led innovation. For example, since R&D involve high fixed costs, they are intrinsically a high-risk activity and are subject to economies of scope that favour larger firms. Other general features of SMEs such as greater vulnerability to the essentially unpredictable market responses to innovative activity, or the greater difficulties they face in accessing financial and human capital, place them in a disadvantaged position with regard to engaging in innovative activities. When it comes to ICT-based innovation, policymakers need to take into account the general difficulties encountered by enterprises in developing countries, particularly in connection with ICT access and use by enterprises.7 This section will consider the ways in which ICTs affect innovation, present examples of how ICTs may contribute to innovation in its different aspects, and explore a number of implications for developing countries.

### How are ICTs Connected with Innovation?

Often ICT adoption by enterprises is regarded as a self-contained event that takes place in the course of an enterprise's development within a given technological, commercial and business environment. From that viewpoint, ICT use in businesses' operations would not be considered fundamentally different from, say, the purchase of more up-to-date machinery. In reality, however, if the full potential of ICT use in enterprises is to be realized, it must be understood and implemented as an integral part of a much broader process of transformation of business structures and processes, a transformation that necessarily involves the emergence of new or significantly improved products or services, and/or the implementation of new productive or managerial methods, namely innovations.

This chapter started with a reference to the value-creation role of knowledge as source of innovation and the central function that the latter performs in the development of a market economy, a view articulated in a path-breaking manner by Schumpeter. However, the knowledge that is available to a society at a given point in time and the knowledge that is actually involved in innovation have not historically been identical magnitudes. Before the explosive development of ICTs, information and knowledge were much more costly to acquire and, in particular, to disseminate. The consequence was that they were much less available to enterprises and society at large, and therefore the pace of innovation was slower: only a few people would have the necessary knowledge, and decision-making processes would be strongly hierarchical (which may have been the most socially efficient solution in an environment of high cost of information). The link between innovation and information remained at the top socioeconomic layers and reached the mass of enterprises only after significant delays. With the fall in the cost of access to information the situation changes radically. The possibility of innovating is open to much wider strata of economic actors. Indeed, for many firms innovation becomes not a possibility, but a necessity. When everyone can innovate everyone must innovate, or risk being undercut by competitors. Abundant information creates a pro-innovation force by itself.

While in an environment of scarce, costly information innovation tended to result more or less exclusively from advances in technology embodied in material capital (better machines), the abundance of information makes innovation in intangibles (for example, brands or business models) increasingly important, hence the importance of innovation in business processes and structures. This is particularly visible in the flattening of organizational hierarchies. As enterprises become more aware of the importance of innovation for their competitiveness, and the need to be proactive innovators, interactions among different categories of employees and managers evolve. Information has to be shared much more widely than in the past and ICT-enabled business processes tend to generate information flows that ignore formal hierarchies.

It is easy, almost trivial, to explain the many ways in which adopting ICTs can improve business performance, particularly by allowing companies to operate differently and at lower costs, and also to bring new products to the market. However, determining with precision what is going to be the net financial impact of particular ICT implementations, choosing the links in the value chain where interventions could yield better results and deciding on the sequencing and pace of adoption are not necessarily equally simple tasks.

Research tends to support the concept that e-business cannot be considered in isolation from other aspects of business change. A common findingis that the positive effects of ICT use can include increased revenue and market share, greater adaptability to market conditions, shorter product development cycles, improved quality of after sales and, as a result, greater profitability. At the same time, research indicates that companies that complement their e-business projects with investments to enhance their workers' skills and to facilitate business process change are the ones that reap most of the benefits of e-business. In the absence of those conditions, the impact of ICT on the performance of enterprises tends to be significantly less.

**Why ICT Still Matters for Innovation?**

It has been argued that ICTs are now so widely prevalent in developed economies that their adoption can no longer be considered to be able to make a noticeable difference in innovation-based competitive strategies. According to this view, ICTs, like other "infrastructure" technologies (such as electricity), would have become essential for business operations and, at the same time, irrelevant in terms of strategic advantage.

Any possible signs of a "commoditization" of information technologies are most likely not applicable to developing countries. Furthermore, even in the context of an advanced economy the effects of ICTs on innovation and hence on competitiveness depend on the extent to which the implementation of those technologies has led to actual changes in business processes. Large variations seem to persist in the modalities by which enterprises incorporate ICTs into their competitive strategies, including innovation-related activities. Another reason for caution is that while the theoretical connections between innovation and ICTs are not difficult to establish, there is some difficulty in gathering hard statistical information about how those two phenomena interact in the real life of enterprises. The statistical measurement of e-business is still in its early stages of development, the notions involved are not always homogeneously understood and technological progress keeps moving the goalposts.

## Changing the Innovation Paradigm

In spite of the above considerations, it can be argued that there are fundamental differences between ICT and other "infrastructure" technologies that make ICT a force for change in production/exchange processes and an accelerator of innovation. In fact, what seems to be happening is that the contribution that ICT makes to the competitiveness of enterprises is not decreasing; on the contrary, the gap between the potential of ICT and the value it delivers to organizations is actually growing. For example, in a recent survey of innovative enterprises in Europe, around half of all innovations introduced in recent years were ICT-enabled, and they seemed to be at least not inferior to other innovations in terms of impact on enterprise profitability. While ICT-enabled innovations were more visible with regard to process innovation compared with product and service innovation, the latter seemed to have a greater impact on profits.

In this context it can be useful to refer to the concept of techno-economic paradigm, according to which a widespread set of related innovations, developed in response to a particular cluster of technical problems and using the same scientific principles and organizational methods, provides the basis for scientific and productive activity at a particular point in time. ICTs have profoundly changed the techno-economic paradigm within which innovation takes place today in developed and more advanced developing countries. Prior to those changes innovation revolved around concepts of mass production, economies of scale and corporate-dominated R&D. In the last three decades of the 20th century this has been replaced to a large extent by an emphasis on economies of scope, exploiting the benefits of interconnected, flexible production facilities and greater flexibility and decentralization of R&D and development. Flexibility, interconnectedness and collaboration rely on ICTs, which also play a fundamental role in facilitating research diversification and collaborative, interdisciplinary approaches.

ICTs enable faster cross-border knowledge dissemination, particularly within transnational corporations, but also by facilitating networking and partnering among smaller players. By investing in ICTs enterprises improve their capacity to combine disparate technologies in new applications. This is important not only from the point of view of ensuring that firms achieve an adequate spread of internal technological undertakings but also from the point of view of the need to engage in R&D partnerships. In that regard, the major benefit of adopting ICTs may not necessarily derive from the technology per se but from its potential to facilitate technological recombination and change.

The relative decline of economies of scale as the massively dominant factor of competitiveness provides another justification for the role of ICT in the emergence of a new innovation paradigm. Product specialization, which through mass production generates competitive advantage, can be complemented in terms of competitive strategies by product or market diversification, which places a premium on the notions of flexibility, customization and adaptability for which ICTs are particularly fitted.

Much as the capital goods sector provided the means by which innovation moved across sectors and industries in earlier innovation waves based on the steam engine and electrification, ICTs are playing a crucial transmission role in the new innovation paradigm. But, instead of being a mere instrument for the transmission of knowledge about how to apply technology to various fields of activity, ICTs facilitate the connection of different fields of innovation potential (inside the firm or among members of value networks). This results in broader scope for innovation.

The dominance of ICTs in the new innovation paradigm also influences the geographical distribution of innovative activities and centres of excellence. Companies that fully understand and implement ICT-enabled strategies can become capable of participating in geographically complex international networks made possible by, among other factors, the new combinations of activities, whose respective centres of excellence may have been sited in distant locations. ICT-intensive innovators can be effective in managing those geographically dispersed networks, and in recombining formerly distinct and distant learning processes. This observation is mostly applicable to the manufactures industries in general and more particularly the ICT equipment sector, in which global learning networks are a common consequence of participation in global networks for the sourcing of technology – a case in point being that of original equipment manufacturers in Taiwan Province of China and in mainland China. In this way, ICTs can be seen as the string that bundles together three trends in innovation in the globalizing economy that were sketched out in the preceding paragraphs:

the emergence of R&D networks that bind specialized poles of innovation, the diversification of innovation activity at the level of the firm, and the emergence of technology-focused inter-firm alliances

## Some Implications for Developing Countries

What does all this mean for developing countries? Enterprises in those countries face particularly difficult conditions in their innovation activities. Major constraints include the following: in most cases they cannot rely on a strong local scientific research community; they have to be able to make a profit while operating in narrow, low-income markets; and they lack the funds to support in-house R&D activities, with little or no venture capital ready to finance innovations. Clearly, ICTs do not provide immediate responses to those problems. However, some of the features of the ICT-led innovation paradigm that were discussed above could help create a more innovation-friendly environment in developing countries.

In the first place, the reduction in the cost of accessing and processing information and knowledge can help reduce the cost of generating the local scientific base of innovation. ICTs' significant enhancement of scientific-content accessibility should not be underestimated. The means for those involved in R&D in locations away from the leading scientific centres to remain abreast of the latest developments in any particular field of study are today vasdy superior and cheaper than they were just two decades ago. In addition to providing easier access to the outputs of scientific research, ICTs are generating, stocking and enabling the manipulation of enormous databases that, by themselves, represent an opportunity for accelerated progress in natural and social sciences. At the same time, one should bear in mind that the generation and the accumulation of knowledge constitute complex social process whose success is not determined merely by the material availability of information. The tacit component of knowledge remains much less amenable to transfer through ICTs than its codified one. History, institutions, and even the physical and geographical conditions in which enterprises in developing countries operate, have a considerable influence on their capacity to move from acquiring information to learning.

Over a shorter time-span, the ICT-facilitated trend towards decentralization and diversification of research and innovation can benefit those developing countries where there is already some innovative activity by facilitating their integration into global learning networks and technology-based alliances. As technological change accelerates, competitive pressures force companies to augment their knowledge and capabilities. An increasingly important way to do that is to use "open innovation", a term that describes how even companies that are leaders in their field must complement their in-house R&D efforts with technologies developed by others, and open up their own technological knowledge to outsiders. Traditional, inward-oriented technological capacity-building efforts tend to be complemented or even replaced with more outward-looking strategies that rely on the output of networks of universities, joint ventures, start-ups, suppliers and even competitors. This often results in technology-based alliances, which are being more and more used as instruments of learning. In addition to providing a fundamental means of cooperation in R&D within alliances, ICTs themselves are among the technologies that are more frequently the subject of technology-based alliances.

Technology-based alliances are increasingly important means of facilitating cross-border technology and know-how transfer. In 2000 estimates of R&D activity that were based on the number of patents indicated that only about 10 per cent of the technology development work of transnational corporations was done away from their home country, and only about 1 per cent was taking place in the developing countries. More recently, UNCTAD indicated that the share of developed countries in R&D expenditure had fallen from 97 per cent in 1991 to 91 per cent in 2002, while that of developing Asia had risen from 2 per cent to 6 per cent. According to Narula and Sadowski, over 93 per cent of the technology partnerships they studied were between developed countries. The partnerships undertaken by firms from developing countries were signed almost exclusively with firms from developed countries. The countries most actively engaged in those agreements are the East Asian newly industrializing countries, together with several economies in Central and Eastern Europe. The participation of African firms was negligible.

On the other hand, UNCTAD found that R&D was internationalizing rapidly and that between 1993 and 2002 the R&D expenditure of foreign affiliates worldwide increased from an estimated $30 billion to $67 billion (or from 10 per cent to 16 per cent of global business R&D). While the rise was relatively modest in developed host countries, it was quite significant in developing countries: the share of foreign affiliates in business R&D in the developing world increased from 2 per cent to 18 per cent between 1996 and 2002.

UNCTAD also indicated that the reported increase in the internationalization of R&D was affecting developing countries in a very uneven way, with developing Asia as the most dynamic recipient by far. The case of the R&D facilities established in India by a number of leading foreign companies suggests some factors that, over and above the lower salaries of scientists and engineers, may account for the decisions to engage in ICT innovation activities in developing countries. Such factors are often considered to include the existence of good-quality scientific and technical universities, which in turn results in the availability of a sufficient number of science and engineering graduates, and the pre-existence of a tissue of commercial and academic entities that are relevant to the relevant area of work.

The interplay of the factors that have made possible the emergence of ICT R&D poles in some developing countries is a complex one. Archibugi and Pietrobelli suggest that the arrival of a leading foreign technology firm may "generate externalities and induce the public sector to give prominence to associated Faculties and other public research centres". They quote as an example the R&D facility established by Texas Instruments in Bangalore as long ago as 1985, which specialized in design circuits. While it is not possible to ascertain whether the specialized ICT hub that currently exists in Bangalore would have developed irrespective of that decision, it seems clear that the large number of ICT firms that have been created and developed there would not have emerged without active public support policies for the ICT sector, particularly those that relate to the generation of a large pool of qualified engineers.

ICTs also facilitate and influence the location decisions in relation to R&D activities of enterprises from developing countries, which may sometimes choose to locate some facilities in developed countries. This may be done in order to facilitate the absorption of best practice and its transfer to the home country production sites. For example, according to Serapio and Dalton, firms from the Republic of Korea owned more R&D facilities in the United States than companies from several developed countries with economies of comparable size. Those facilities were mainly operating in the field of ICTs. More recently, UNCTAD reported that the foreign R&D activities of developing country TNCs were growing rapidly. That trend is driven by the need to access advanced technologies and to adapt products to major export markets. Some of those TNCs are targeting the knowledge base of developed countries, while others are setting up R&D units in other developing economies.

**Open Innovation Approaches**

Open source software represents an innovation-oriented form of ICT-focused collaboration that is of increasing interest to enterprises in developing countries. In this case, ICTs provide not only the origin of the conceptual framework for a new mode of production and dissemination of knowledge and innovation, but also the main tools through which that framework can be put into operation.

Open source software, which consists of collaborative projects involving people located around the world and who, had it not been for the Internet, would never have had an opportunity to pool their knowledge, is one of the major examples of Internet-driven social and economic change. Free and open source software (FOSS) provides the most visible model of that trend. A discussion of the economic rationale for the open source software model and the advantages it offers from the point of view of developing countries is to be found in UNCTAD.

With regard to innovation, FOSS gives developing countries the opportunity to benefit in a number of ways. In the first place, thanks to FOSS, software developers are able to enhance their skills and to improve more easily on ICT (software)-based business models. Participation in FOSS projects facilitates the creation and consolidation of local pools of expertise in ICTs, and given the low barriers to access to that software, enables the emergence of new ICT-related business activities. This may result in the launching of new (innovative) products or services, geared to the specific needs of local demand, that may not have been possible with proprietary software.

FOSS also reduces the cost of access to ICTs, although it should be stressed that cost considerations are not the primary advantage of the FOSS model for developing countries. Since ICTs, as discussed above, are powerful enablers and accelerators of innovation, FOSS can play a crucial role in improving innovation in developing countries by supporting the accumulation and dissemination of knowledge. As FOSS uses open standards, it provides all economic actors with access to this key technology on equal terms. This works against the emergence of monopolies, favours technological diversity and promotes the emergence of new innovative business models.

The approach followed in the implementation of open source software projects is now being applied in fields outside software development. In the case of open source software, individuals cooperate on line with their peers in the production of software that is openly available to anyone who is interested in carrying the work further. The same model of on line collaboration can be applied to any business model in which the end product consists of information.

The open approach to innovation is based on the notion that as knowledge becomes more widely available and increasingly varied in sources and content, companies cannot limit themselves to their own research in order to be innovative and stay competitive. They have to be able to integrate and exploit external knowledge and skills. ICTs are used to integrate the originally highly dispersed information and capabilities in a trial process from which innovation emerges. This is as valid for the individual enterprise as for cooperative networks of enterprises organized online. This open cooperation process of production can take place on a volunteer basis, with the end product being made available to everyone at no cost (such as Wikipedia). Alternatively, some form of compensation may be demanded by contributors and the end product may be used in a commercial operation.

Open innovation is related to the idea that competent users should be integrated into the innovation process. This refers to the trend for users (whether enterprises or consumers) to develop themselves the product or service they need and then freely release their innovation, as opposed to the traditional, manufacturer-centred approach whereby a manufacturer identifies the users' needs, develops the products and profits from selling them. User innovation is common in many fields, from scientific instruments to computer applications (for example, e-mail) or sports equipment. Users tend to innovate collaboratively in communities, a process rendered much more effective by the Internet. They also tend to grant free access to their innovation because often it does not make sense to protect it. Their major motivation for innovating is the improved usefulness of the product, which can be enhanced even further if other users have access to it so that they can improve on it. In many cases, their reputation as a skilled user-developer also matters strongly to them. The innovativeness of the results of the process can be so great that some suppliers may even decide to concentrate exclusively on manufacturing, giving up all product development and limiting themselves to downloading user-developed designs from user community websites. As an example, von Hippel mentions cases in fields as different as kitesurfing equipment, spine-surgery devices and toys. It is clear that the social benefits of user-led innovation would be significantly reduced without ICTs that provide users with strong computing capabilities needed for product design at a reasonable cost, as well as fast, cheap and far-reaching means for collaboration in users' communities.

The growing reliance by online media on "user-generated content" and the emerging phenomenon of "crowdsourcing", whereby the Internet is used to involve interested individuals in the definition and development of new products or in the identification of a solution for a particular problem, are similar manifestations of the capacity of ICTs to democratize innovation processes.

Other business models have been developed for using the Internet to harness the resources of large pools of individuals or firms in order to address innovation problems in an open context. This includes NineSigma. com, which claims to have access to a global network of 1.5 million experts. According to the company, 60 per cent of the experts come from companies, 30 per cent are academic researchers and the remaining 10 per cent are with research laboratories, either publicly or industry-funded. Since those experts are free to forward the request for a solution to an innovation problem to other experts they know, as much as 40 per cent of solutions provided come from experts that were not originally contacted by NineSigma. The company reports that solutions have been provided by sources in a number of developing and transition countries, including Bulgaria, China, India, the Russian Federation, the Syrian Arab Republic and Yemen.

Another example is Yourencore.com, which matches demand from companies looking for expertise regarding a particular scientific or engineering problem with supply of expertise from retired scientists and engineers who can thus earn extra income while working according to their own flexible schedule and staying intellectually active in their preferred professional field.

Based in India, Ideawicket.com has recently launched an "Open Innovation Portal", which provides a platform for individual innovators and corporations to exchange their innovation resources and requirements. Innovators are also able to give visibility to their creations and eventually make them available to interested enterprises.

A similar concept with a developing country involvement is the OpenBusiness project launched jointly by Creative Commons UK, Creative Commons South Africa and the Centre for Technology and Society at the FGV Law School in Rio de Janeiro, Brazil. They describe OpenBusiness as a "platform to share and develop innovative Open Business ideas-entrepreneurial ideas which are built around openness, free services and free access".

Open approaches to innovation can take different forms. In some cases they may consist in a form of outsourcing of part or all of a company's R&D. Contracts and market conditions are the fundamental factors governing such projects. ICTs are helpful, but not necessarily crucial. In other cases, enterprises may go for a closer cooperative arrangement where reciprocity, long-term strategic considerations and partnership are the keys. ICT-enabled cooperation becomes an important element in that case. Finally, taking the approach one step further they can go for what one could describe as "open access" innovation, along the conceptual lines of the FOSS model. In that case, innovation is the work of networks involving enterprises, users and eventually other stakeholders. Transparency, reputation and trust are the keystone concepts and ICTs are indispensable for the implementation of such projects.

The emergence of an ICT-based paradigm of innovation makes it advisable to gear government intervention to the promotion of cross-firm and cross-border knowledge flows, it being assumed that firms follow the model of a continually interactive search for better methods and improved products, and hence a Schumpeterian search for higher profits through experimental innovation. This could be more socially efficient, particularly from the standpoint of developing countries than the more traditional approach of protecting the exclusive right of the private owners of the resources invested in the generation of knowledge to protect that knowledge. The next section looks at how Governments can shape their e-business and ICT policies so that their interaction results in faster, more development-relevant innovation.

## Maximizing Synergies between E-business and Innovation Policies

While e-business is approaching maturity status as an operational technique and there is no controversy about the role of ICTs in recent business innovation waves, the recognition at the policy level of the need to simultaneously consider e-business and innovation issues is recent and incomplete, and to a large extent led by developed countries. As developing countries adapt their national innovation systems to benefit from the dynamic interplay between ICT use by enterprises and innovation-led competitiveness policies they need to be aware of available experience in this regard and adapt the lessons from it to their specific needs and concerns. This section will present the main lines of convergence of e-business and innovation policies, and identify best-practice examples that may be relevant and transferable to developing countries.

## Emerging Approaches to Innovation Policy

Innovation policy aims at increasing the amount and enhancing the effectiveness of innovative activity by enterprises. As discussed above, broadly defined innovative activity not only includes the creation, adaptation or adoption of new or improved products or services, but also refers to many other value-creation processes. Therefore, innovation policy needs to consider not only strictly technological innovation but also organizational (new working methods) and presentational (design and marketing) innovation. Given the strong links between ICT use by enterprises, competitiveness and innovation, it is clear that enterprises need to align policies in those fields.

In practice, the integration of policies to promote ICT use by enterprises within general innovation policies still has some way to go in most countries. Two broad categories of approach can be identified in this regard. A relatively small number of countries, mostly in Northern Europe, have adopted an approach to innovation policy that sees it as a cross-cutting issue that must be included in the policies of individual agencies, together with ICT and e-business policies. This often leads to a reduction in the number of independent initiatives undertaken by individual agencies and ministries. A second, more common approach still sees innovation as a dual phenomenon. On the one hand, it is perceived as being the result of R&D processes, support for which is in the hands of agencies dealing with research and education. On the other hand, it is also considered to be a tool for the enhancement of competitiveness and SME modernization and this aspect of innovation is handled by ministries in the sphere of economics, industry and trade. In this approach, programmes to support innovation may include e-business as one issue among others to be dealt with. The instruments used, however, may not be too different from the ones chosen by policymakers who focus primarily on e-business as an innovation driver. This is particularly true in the case of SME-targeting programmes, since for this group of enterprises adapting to the changes in the competitive environment that have been brought about by e-business is probably the priority innovation strategy.

While the links between the policy framework of innovation and that of e-business policymaking are becoming stronger in a growing number of countries, the range of policy interventions and actors remains wide, particularly since different levels of Governments and quasi-governmental agencies see innovation, competitiveness, e-business and/or entrepreneurship support as crucial to their development objectives. From the substantive point of view, the differentiation between ICT/e-business policies on the one hand and innovation policies on the other remains visible with regard to several important dimensions of the problems in question. For instance, standard innovation indicators are not frequently considered relevant to efforts to benchmark the development of the "information society". On the other hand, it is common to see e-business promotion plans that lack an explicit reference to innovation as an objective, even though different forms of innovation represent the core goal of those plans. In practice, this results in e-business measures that are implemented at the operational level by the same ministries or agencies as innovation measures, ones that address the same target population and share policy instruments, but lack a clear common conceptual framework and formal feedback mechanisms, and waste opportunities for synergies.

The importance of integration and coordination of policies from different ministries and at different levels must be stressed. Since e-business affects many sectors and processes, an integrated approach to the issues is more likely to produce practical results than a piecemeal one. A way to seek coordination can be to bind together all instruments for ICT use promotion and innovation support in one single public sector ministry. However, that does not always produce optimal results in terms of integration. A way to improve that is to bring together in policy formulation all different stakeholders. Up to three levels of policymaking can be distinguished in this field. One includes decision makers in national ministries and agencies. A second level is formed by entities such as internal agency working groups, interagency commissions and the like. The third level involves different ICT user groups. While the clear distinction between the policy and the operational levels should be maintained, the involvement of all those stakeholders is necessary, and there should be channels to ensure a reasonable balance between top-down and bottom-up policymaking. It is also important to support networking across sectors and across innovation themes.

**Adapting Best Practice in E-business and Innovation Promotion**

There is a major opportunity for developing countries to learn from best practice in the field of public support related to e-business and innovation. They should explore some of the existing mechanisms of governance and weigh up the pros and cons of pursuing national, regional and local agendas and also exploring alternative ways to distribute public, private or public/ private responsibilities for implementation.

The first issue to consider is the integration of the institutional frameworks of innovation and e-business policymaking. In that regard, many of the developed countries, particularly in Europe, have entrusted overall policymaking for innovation and e-business to the same organizations. However, this does not imply that policies are necessarily coordinated. Interestingly, when one considers the case of the European Union, it is noticeable that among the newer member States the dominant trend is to give institutional pre-eminence (in the form dedicated ministries or commissions) to the issues of ICTs, the Information Society or e-business, while in the older member States - the EU 15 - those matters are more likely to be placed in the hands of the institutions generally in charge of innovation, scientific R&D and industry.

Norway and Iceland are examples of countries in which two different ministries are involved in those matters (in Norway the ministries of trade and industry and of education and research; in Iceland the ministries of education, science and culture and of industry and commerce). In other countries innovation and e-business policies are addressed in parallel. This is the case of the Netherlands, where the Ministry of Economic Affairs has responsibility for both policy types, but e-business issues are delegated to an external agency, while other ministries are also involved. This does not, however, ensure policy coordination and the linkage of innovation and e-business policies. As a response to the challenge of integrated governance an Interdepartmental Committee for Science, Innovation and Informatics (CWTI) has been to created to prepare the overall policy strategies on science, research, technology and innovation policy.

An example that has been regarded as representing best practice, addressing both the innovation and the ICT policy areas is the Italian Action Plan for ICT Innovation in Enterprises, which was launched in 2003 by the Ministry for Innovation and Technologies and the Mnistry for Productive Activities. In addition to questions of ICT and innovation, the Plan dealt with other elements that play a role in improving the development potential of enterprises, such as education and training, R&D and entrepreneurship. It also included a consultative e-business committee composed of representatives of academia, business associations, financial institutions and trade unions.

Malaysia, a leading performer in terms of innovation and e-business adoption among developing countries, provides an example of efforts to integrate ICT and innovation policymaking. ICT policy-making was transferred from the former Mnistry of Energy, Communication and Multimedia to the Mnistry of Science, Technology and Innovation (MOSTI). MOSTI's mission statement reads as follows: "harnessing Science, Technology and Innovation (STI) and human capital to value-add the agricultural and industrial sectors and to develop the new economy, particularly through information and communications technology (ICT), and biotechnology". Thus, ICT policy is conceived as an integral part of STI policies. Among other aspects of ICT policymaking, MOSTI is responsible for the formulation and implementation of national policy on ICT and the encouragement of R&D and commercialization in ICT.

The case of Mexico illustrates a different approach to the relationship between innovation and ICT policymaking. In response to strong competition from other developing countries, Mexico has undertaken a number of initiatives to upgrade the innovative potential of its economy. Many of those initiatives to expand its science and technology base are undertaken under the leadership of the National Council for Science and Technology (CONACYT), an agency that reports directly to the President of the Republic. Other government agencies are also implementing programmes to enhance the productivity and competitiveness of the private sector, enhancement being considered one of the strategic objectives of the national development plan. Several of those programmes relate to ICT and e-business issues. There are several other ministries and agencies, in addition to CONACYT, that have responsibilities connected to ICT policies. They include the Mnistry of the Economy, the Mnistry of Communications and

Transport, the Mnistry of Education, the development bank and the 32 State governments.

The role of the State governments in the Mexican case is an example of a situation that is found across developing and developed countries. Regional and local governments play an important role in implementing innovation and e-business policies, although in most countries the largest role in policy definition and coordination remains at the national level. Regional and local levels of policy delivery often seem to be highly relevant, as in most countries most enterprises are of local or regional size and only relatively few operate at the national level. Programmes commonly rely on local or regional networks of stakeholders or communities of practice for the delivery of support measures. While many factors influence the ability of firms to innovate, some of the most direct ones are often not easily handled at the national level. Clusters, for example, are a powerful instrument in innovation policies which is normally managed at the regional level. This does not mean that there is no role for policy at higher levels of aggregation: the operation of clusters can be improved through networking, the identification of best practice and common learning. National frameworks provide useful support for those aspects of innovation and e-business policies.

### Major Support Instruments for E-business Innovation

Innovation takes place primarily within firms. The priority concern of Governments therefore is to establish and implement sound innovation policies that enable firms to maximize innovation. Creating an environment that facilitates innovation often requires the addressing of a number of market failures that may result in suboptimal levels of innovative activity by the private sector. Those failures may relate to the operation of markets (financial or goods and services) or limit the presence of knowledge spillovers. Instruments commonly used to address such problems include the direct funding and subsidizing of R&D activities, the provision of support for venture capital funds, the facilitation of transfer of technology and partnership networking mechanisms, incubators and clusters, demand-side management, the provision of data, analysis and studies, and the establishment of centres for demonstration and testing. E-business policy interventions are very frequently present in most if not all of those categories of instruments. For example, in R&D funding, public funding is frequently granted for research in areas such as software, expert systems or Internet technologies. E-business is also frequently chosen as a field in which multi-stakeholder partnerships involving enterprises, business associations, educational institutions and government agencies aim at facilitating transfer of technology and innovation.

Funding of R&D activities tends to be the innovation policy instrument most frequently used. Technology transfer, partnerships and networking are other policy tools that are frequently used. Frequently, programmes supporting e-business and innovation cover more than one policy area in scope. Projects and funding for the support of feasibility studies are also well represented, possibly as a consequence of the ongoing experimental nature of e-business and e-commerce development. Box 4.3 presents some examples of specific programmes that link e-business and innovation.

In some cases policy interventions do not establish a clear-cut distinction between e-business goals and broader societal objectives such as the improvement of e-skills, e-government initiatives or ICT infrastructure and access issues. In that context, e-business is sometimes perceived not so much as a policy area in itself but as rather an instrument to achieve goals in respect of a number of Information Society issues.

From the point of view of the relationship with innovation policies, the traditional approach still tends to see e-business more as part of the ICT policy field than as a component of the innovation policy set. One reason for this is that ICT has been a major force for change in many policy areas beyond innovation (health, employment, security, education and so forth). In practice, however, the application of policies that may a priori be categorized as coming within the scope of either innovation or e-business policy often results in the development of responses to economic problems that involve a subject crossing from one policy field to the other. It is increasingly difficult to establish a clear-cut distinction between those two cross-sectoral groups of policies. As empirical evidence for the impact of ICT and e-business on economic performance mounts, the logic of reinforcing innovation policy by incorporating e-business and ICT support policies becomes stronger.

While it can be expected that e-business and innovation policies will tend to become more closely integrated in the medium and long term, it also clear that the connection between e-business policies and general ICT policies will remain stronger in many countries for some time. A reason for this is that the implementation of ICT and e-business investment by a company or government agency may not translate into an immediately noticeable improvement in their capacity to innovate. In fact, the potential to transform business operations may not even be the explicit policy objective of ICT adoption, although in the end their application may generate very significant innovations. Cost-cutting, improving the relationship with customers or many other straightforward business needs may be the explicit reasons for e-business adoption. Sometimes the objective can be of a more strategic nature, such as driving and facilitating profound changes in the organization.

### Getting the SMEs on Board

Given the huge weight of SMEs in developing economies, particularly from the point of view of employment, the question of how to address their particular needs in the field of ICT and innovation should be carefully addressed.

The first point to consider is the way in which innovation and e-business affect the competitive and innovative behaviour of SMEs. In that respect it is important that policymakers keep in mind the need to adapt their instruments as ICT-driven innovation spreads in the economy and creates network business models that may affect the relevance of policies and interventions. A related business environment factor to consider in the design of innovation support programmes targeting SMEs is that as e-business practices generate, SMEs find that innovation becomes more easily visible to, and eventually imitable by, competitors.

Another consideration that is specific to SMEs and may require special efforts from innovation support policies concerns the need to make clear to the SMEs the relevant benefits that a particular intervention delivers to them and the specific manner in which their competitive potential will be enhanced as a result of it. The amounts involved may be significant.

Helping SMEs identify how innovating through e-business can be actually done in their particular environment can be a time-consuming and expensive exercise that needs to be conducted in an individualized manner. Policies should aim at helping SMEs integrate ICT and e-business considerations as a fundamental element of their enterprise development plans. This means that programmes need to focus on what ICT and e-business can do to solve real problems that SME managers can easily identify. An adequate outreach strategy in that regard needs to make extensive use of the language that enterprises understand best: the financial performance benchmarks that they are used to. When SMEs can make a clear connection between their performance benchmarks against those of their competitors and their relative position in terms of ICT, e-business use and innovation the vital importance of the questions involved comes across to managers immediately.

An important lesson that countries which are considering putting in place support programmes in this field need to keep in mind is that for initiatives to succeed they need to remain in place for a reasonable period of time. The value of any set of ICT innovation support measures can be judged on a rational basis only once some impact measurement has been undertaken, and this takes time. However, it is not uncommon that programmes in this field are terminated before their effects on enterprises can be assessed. This makes it difficult to replicate and scale up successful initiatives, and to accumulate and disseminate best practice. At the same time, it is also important that policies adapt and change in response to practical experience. Striking the right balance between the needs for policy stability and flexibility and evolution calls for mechanisms that allow feedback from end-users to reach policymakers, and frequent and meaningful interaction between all stakeholders.

## Policy Recommendations

An innovation policy framework that fully takes into consideration the changes generated by ICT must give prominence to open approaches to innovation, which present significant advantages for developing countries. In order to support open innovation, which relies on cooperation, policymakers need to take into account the trade-offs that exist between the competitive incentives to innovate and the ability to cooperate with other enterprises. Partnerships and alliances, between the private and the public sector and between enterprises, become particularly useful in that regard and ICT provides platforms that make them easier to implement.

An open innovation approach requires some adaptation of innovation support policies. The emphasis should move from supporting individual firms to providing more assistance to innovation networks. Many funding or incentive instruments to support innovation currently orient firms towards building closed, in-house R&D capacity and do not encourage them to look for networking opportunities. In many countries, enterprises that rely on cooperation with external networks in order to innovate are not eligible for innovation support measures, which are geared to firms with internal R&D facilities only. Supporting networks of innovation also involves strengthening networking skills, particularly among SMEs; this includes aspects such as human capital management, intellectual property management and trust-building

Public research institutions and academic centres have a key role to play in advancing open innovation. The academic incentive system may benefit from reconsidering how to stimulate the emergence of collaborative strategies, to foster knowledge-sharing and to engage with industry.

User-driven innovation is a form of ICT-enabled open innovation that is increasingly important. Policy can support it in a number of ways. One is to promote the use of open standards, particularly through public procurement policies. Another one is to undertake efforts to assess the extent to which user innovation is actually implemented in order to adapt overall innovation policy. Policymakers should also consider the possible negative effects on user-driven innovation that may result from imposing restrictions on the ability of users to modify products that they own.

Finally, general environment conditions that are frequently mentioned as necessary for the development of an information society are also important for the development of open innovation. These include an educated population, IT infrastructure of good quality, efficient capital markets and the creation of a high-trust business environment.

As the effects of e-business adoption spread across the economy and affect most sectors and aspects of business process an integrated approach to e-business and innovation policies is needed. Experience shows that the effectiveness of innovation policy and e-business-related policies improves when they are conceived and implemented in a closely coordinated manner. But the changes that ICT have introduced in the functioning of modern economic systems are such that this coordination between both policy areas is likely to gradually lead to a significant level of integration among them.

In practice, the integration and coordination of policies in those two fields amount to the creation of an institutional framework of cooperation among the different ministries and different levels of government that are often involved in them. Coordination can be achieved by integrating all instruments for e-business and innovation support into a single ministry or agency, although this does not always result in closer coordination. It can also be achieved by creating institutional mechanisms that allow the full involvement of all stakeholders, with the right balance between bottom-up and top-down policymaking.

Enough experience of policy approaches and instruments is already available to suggest that for developing countries ICT-enabled innovation policies provide an opportunity to explore open innovation approaches that may be more suited to their concerns than traditional ones. Sharing of experiences and learning to learn from others will be crucial in that regard. This is a process that UNCTAD may be well placed to facilitate at the global level as part of its work in the field of science, technology and innovation, in cooperation with other relevant organizations at the regional level.

## References

Archibugi D and Iammarino S. (2000). Innovation and globalisation: Evidence and implications. In Chesnais F, Ietto-Gilles G and Simonetti R (eds.), *European Integration and Global Technology Strategies*. Routledge. London.

Cantwell J A. (1999). Innovation as the principal source of growth in the global economy. In: Archibugi D, Howells J and Michie J (eds.), *Innovation Policy in a Global Economy.* Cambridge: Cambridge University Press.

Cassiman B and Veugelers R. (2004). Foreign subsidiaries as channel of international technology diffusion: some direct firm level evidence from Belgium. *European Economic Review,* 48:2, pp. 455-76.

Criscuolo C, Haskel J and Slaughter M. (2005). Global engagement and the innovation activities of firms. N*ational Bureau of Economic Research Working Paper* No. 11479.

Narula R and Sadowski M. (2002). Technological catch-up and strategic technology partnering in developing countries. *International Journal of Technology Management,* vol. 23, no. 6.

# 9

# E-Banking and E-Payments

Financial service providers, and especially banks, had been using electronic messages for quite some time before the introduction of the Internet. Those mes-sages were transmitted through proprietary software systems, also known as intranets, through modems linked to clients' personal computers or phone lines permitting them to consult accounts and make chang-es in payment orders in an offline regime. The Inter-net revolution started the process of replacing those traditional methods of electronic communications by Internet Protocol (IP) based systems. The literature is increasingly focusing attention on this newer and more widespread development of e-banking and e-payments, while continuing to discuss also the role of payment cards, automated teller machines (ATMs), tel-ephone banking and mobile banking (m-banking) or m-payments. The latter can use IP and other communication protocols and are relatively more important in the context of developing countries.

The use of modern information and communication technologies (ICTs), and especially the Internet, by financial services providers greatly increased their communications capacities and speed, decreased the transaction costs of financial operations and permitted the networking of a host of players in various financial schemes. As a result, it brought about major business process innovations. For example, ICTs accommodated the explosion of large-value international payments traffic especially during last two decades, in particular thanks to the introduction of new online payments protocols and real-time gross settlement (RTGS) systems.

E-banking and e-payments, both corporate and retail, proved to be less cosdy for the commercial banks and at the same time more convenient for businesses, Governments and households. The use of e-finance relies on bank deposits and diminishes the role of cash money (notes and coins). However, it has created another set of security challenges such as the need for protection against emerging cybercrime, which has introduced further innovations allowing more secure methods of e-banking and e-payments.

The intensive use of ICTs also facilitated the transfor-mation of traditional bank-related debt such as various loans and mortgages into securities circulating in capi-tal markets. That increased trading and securitization activities of banks coincided with their relatively re-duced role as deposit-taking institutions, characterized by long-term relations with their clientele.

Banks and payment card providers remain at the core of e-banking and e-payments. However, relatively new players such as non-bank money transfer operators, mobile phone operators and e-payment technology vendors are also coming into the picture, trying to carve out niches or special value-added operations from the main players or to conclude various cooperative arrangements with them.

The financial flows between developed and developing countries are also increasingly taking place in the framework of major online inter-bank transfer systems. While those systems are facilitating the transmission of the main private and public finance flows such as bank credits, foreign direct investment (FDI), portfolio investments and official development assistance (ODA), ICTs are no less important for retail and small-volume financial transfers to households and small businesses in those countries.

The most important small-scale private financial transfers are migrant remittances, which are increasingly relying on online money transfer systems, with a consequent saving of resources for both originators and end-users of those funds. Since remittances have considerably exceeded ODA in their overall volume and are currendy the largest non-debt-creating financial flows, the prospects for the moving of this major source of development finance towards electronic channels constitute an important element in analysing the implications of e-finance for developing countries.

Making e-banking and e-payments more affordable to banks and their clients in developing countries (and to less affluent parts of the population in developed countries) is still a major issue. Furthermore, giving to small and medium-sized enterprises (SMEs), micro-enterprises and individuals (part of whom are "unbanked", i.e. have no bank accounts) better access to simple forms of e-banking and e-payments or inpayments is also becoming an increasingly major challenge, which has started to be addressed very recently.

While acquiring knowledge of modern e-banking and e-payments techniques is still the main hurdle for developing economies, the lack of deeply rooted proprietary systems of finance might become a positive factor and enable countries that are not really attached to traditional systems to leap frog to e-finance. Exploiting that potential will require that the financial sector of developing and transition economies have the capacity to move rapidly towards modern ICT-based systems.

## ICTs and Innovations in Banking and Payments

The active use of ICTs during the last two decades further released the forces of globalization, liberali-zation and technological change. ICTs were particularly driving major innovations in business processes, and especially in financial intermediation. One of the most innovative industries in terms of using ICTs, the financial sector, increased its share of value added particularly in developed economies. For example, in 2006 its share in the United States GDP exceeded 7 per cent, that is it more than doubled during the last 50 years.

Another indicator of the so-called financial deepening of the economy is the ratio of financial assets (bank deposits, government debt securities, private debt securities, equity) to the economy. The volume of financial assets currently exceeds gross domestic product (GDP) by a factor of 3 to 4 in the main developed market economies. Financial deepening is also becoming more and more evident in China and other emerging Asian economies (Indonesia, Malaysia, Philippines, Republic of Korea and Thailand), where it represents double the local GDP. The same process is still lagging behind in Latin America (except Chile), Eastern Europe and especially Africa.

Meanwhile, the use of the Internet as an increasingly important channel of distribution for financial services, and, in particular, e-banking and e-payments, was the most important innovation making it possible to move liquidity and various financial instruments online at much lower transaction costs and higher speed, with a consequent change in the nature of monetary transmission and financial intermediation. The possibility of intensive use of ICTs made it possible also to improve transparency in financial operations. As a result information asymmetry in financial markets was reduced and they were made more complete. In addition to the migration of traditional banking and payments to the Internet, the use of ICTs facilitated the large-scale introduction of securitized forms of traditionally bank-based debt instruments, thus allowing capital to be allocated at lower cost and with a better risk profile. Active participation in that process by insurance companies, pension funds, private equity funds and hedge funds was also supported by the use of ICTs. Those major changes provided not only new opportunities, but also created challenges, the greatest of which was the problem of ensuring the security of online financial operations. To respond to security threats, financial service providers had also to introduce innovative solutions to protect their clients against inventive cybercriminals.

The manifestations of financial innovation are having an increasingly direct impact on developing countries, which are not overburdened by older technologies and can more easily start their e-banking and e-payments operations in a new environment and with new technologies.

## Emerging E-banking, E-payments and E-money

In the heyday of the Internet revolution expectations regarding the development of e-banking and e-pay-ments were highly optimistic, and not without reason. It was a trial-and-error process, with some models be-ing increasingly marginalized or even destroyed during this major and mainly successful process of creation. Thus, models that did not survive or keep pace with the dynamics of overall trends included the majority of new purely Internet-based banks and e-cash ventures. At the same time, responding to the challenges of In-ternet-only banks, nearly all major traditional banks introduced their e-banking operations starting, with moving online the inter-bank payments traffic and supporting online the operations of their clients. With time, major banks expanded their e-banking activities, in many cases at the expense of traditional banking ones (known as "brick and mortar"), and created a twin model also known as "click and mortar" or "brick and click", with the "click" part profoundly innovat-ing and still continuing to transform the banks' overall business models.

The introduction of new technologies especially in the initial stages, involves major investments and sunk costs that can be recovered with increasing economy of scale. Research comparing Internet banking experience with the experience of old and newly chartered traditional banks in the United States, during the period 1998-2003 showed that Internet-only banks had a lower return on assets (ROA), a lower return on equity (ROE) and lower net interest margins, and had to offset an inferior accounting performance, higher loan losses and higher borrowed fund costs with higher capital ratios, even in comparison with newly chartered traditional banks. The authors assumed that Internet-only banks can become successful competitors if they are able to improve economies of scale, gain adequate market share and combine new products with traditional banking products.

The Internet as a financial services delivery channel is acquiring its market share mainly through innovations in traditional banks' organizational structures. The integration of new Internet technologies also required training and retraining of employees such as account officers, integrating new "front office" remote services into the "back office" services menu, moving towards holistic and standardized IT systems, and making the necessary changes in the process of communications and management.

E-banking delivery channels for both major and retail clients are based more or less on the same principles of online operations with clients' accounts, including account consultation, transfers, the use of digitized giros, bills, cheques and other forms of payment, and placing orders for stocks, bonds, derivatives and other securities. At the same time regrouping these and other orders and channelling them between banks generate a huge payments traffic that is handled by major interbank payments systems.

The automation of cashless payments and the dramatic reduction in their transmission costs were ICTs' main contribution in the payments systems. Moving inter-bank payments traffic to the Internet was also a formidable innovation challenge, one which the international banking community has been addressing, particularly during the current decade. That in turn brought about an explosion in the value of payments traffic as a result of financial innovations. Thus, during the last 25 years the value of payments traffic in G10 countries increased 13-fold, and its volume threefold. A more rapid increase in larger value payments, which in many major economies accounted for three quarters of GDP, explains the predominance of those payments over retail payments, with the latter representing only a very small percentage. A major innovation here was the move from deferred net settlements (DNS) systems, whereby the settlements were made by the end of the day, to real-time gross setdement (RTGS) systems. The possibility of settling in the realtime regime was a major step forward in managing systemic risks related to the time needed to clear and settle the inter-bank payments.

In retail e-banking and e-payments there were two main trends in the introduction of innovative solutions: the modification of traditional payments methods such as credit transfers, direct debits, card-based payments and their adaptation to various IP-based applications, and the development of such relatively new instruments as e-money and other pre-funded payments, cumulative collection or integrated payments, payment portals, and (becoming increasingly important) mobile payments. One of the results of the expansion of e-banking and e-payments was substantially lower costs related to initiation and handling of paper orders. The introduction of various e-payments methods decreased in G10 countries the share of payments by paper-based cheques and credit transfer orders from 60 per cent in 1989 to 20 per cent in 2005.

Given the relatively greater importance of mobile telephony in developing and transition economies, it is important to stress the role of m-banking and inpayments as a major phenomenon in e-payments-related innovations. The possibility of using cell phones and other electronic devices as prepaid payment cards was a major step forward in mainstreaming poor and "unbanked" population into e-payments. In the case of m-payments the central role is mainly played by mobile network operators (MNO), while banks are backstopping them by keeping the consolidated accounts and the float accumulated by those mobile operators.

One of the key functions of e-payments is to support B2B, B2C, P2P and other e-transactions. At the same time e-payments can also service offline business transactions that accept deferred payments.

The introduction of e-banking and e-payments has decreased the tendency of economic agents to keep cash, namely notes and coins issued by Central Banks (CBs), and increased the importance of commercial bank money. Indeed, the value of deposits kept by non-banks in commercial banks in G10 countries increased from 30 to 50 per cent of GDP, while the share of currency remained stable at 7 per cent. At the same time, many attempts were made, especially at the height of the Internet revolution, to replace cash with various types of e-money.

Understanding innovation related to e-banking, e-payments and e-money will be incomplete without looking at how they impact on monetary circulation and transmission. Thanks to the Internet, e-banking and e-payments are increasing the speed and diminishing the costs of financial transactions such as payments, as compared with the similar operations intermediated by other mediums such as cash or traditional bank transfer. Because of the greater speed of transfers, e-banking and e-payments are also increasing the velocity of money and hence decreasing the demand for money in circulation at a given level of economic activities.

While e-banking and e-payments are becoming more important in the operations of financial intermediaries, they also have to cope with the inertia of habits and the preferences of retail and corporate customers while servicing their needs. Those preferences are determined by many economic factors, including the propensity of many households and SMEs to engage in cash transactions in order to preserve the anonymity of their transactions, especially in the context of the informal economy. In such a situation e-money can serve as a medium for a particular part of a transaction requiring an electronic transfer of funds, which then might be again transformed into cash. Also, money might be transferred through a sequence of electronic transfers, part of which are not Internet-based and hence are more expensive. In other words, since they are the most efficient way of effecting payments and other financial operations, the IP-based e-banking and e-payment systems can either replace more traditional and more expensive ones, or coexist with them according to the preferences of monetary circulation participants, including those in the "unbanked" population.

There were also attempts to replace currency (notes and coins in circulation) issued and backed by CBs with electronic carriers of anonymous private e-monies as a means of payment embodied in easily portable and mobile electronic devices. In that connection, it is important to understand that private financial intermediation does not really bring about the creation of e-monies that coexist or even replace public e-money backed by a CB and used by banks in their online transactions. In fact, the Bank for International Setdements (BIS) uses the narrow concept of e-money, describing it as a "stored value or prepaid product in which a record of the funds or value available to the consumer for multipurpose use is stored on an electronic device in the consumer's possession". In other words, it is a transfer of value originated either from fiat money (cash) or from bank accounts into an electronic device embedded in a card, or a purse, or software linked to a PC. Having been transferred it becomes e-money making possible further payments. As those devices are produced by private entities, regulatory issues of ensuring their usability at face value may arise. Also, most of e-money systems are not anonymous – in other words the convenience of having e-money comes at a price. As a result, the volumes of e-money are still very small and this type of innovation is still at a very early stage compared with, for example, Internet banking or m-payments.

The proliferation of parallel trusted systems made it possible to consolidate payments, clear mutual debts of, for example, auction participants and settle the remaining debts at the end of regular periods. In some private exchange systems, private e-monies are usable between participants, but unlike currencies are not legal tender. Such mutual debt-clearing systems should normally reduce the demand for money in circulation. The non-bank electronic billing and clearing systems are acting as consolidators of mutual claims between participants of a given network and are then netting out the residual debts in the system by using cash and bank accounts, in other words generally accepted means of payment.

So far, the expectation that e-money would spell the end of notes and coins has proved to be premature. While some elements of those processes are still at an early stage, e-payments, especially in the retail sector, have continued to be centred on payment cards, which have been increasingly becoming smart cards adapted for the Internet. Meanwhile, the expectation that payment cards would be all the more important by now has also proved to be exaggerated.

To assess the role of e-banking and e-payments in monetary circulation it is important to understand which part of payments and other transfers, intermediated mainly by commercial banks, is passing through the Internet and thus enjoying the low transaction costs and speed of that huge network. In other words such elements of Ml as coins and notes, could become, sight deposits and e-money, that is be easily transformed through online banking and payment channels into their purely electronic forms and then revert to more traditional forms based on the preferences of the beneficiaries of those funds.

The modernization of payments systems that rely to a considerable extent on the availability of low-cost, real-time online communications and networking requires better regulation of payments systems in various countries and its coordination, and closer links between regulators and financial service providers in order to rapidly overcome problems due to possible disruptions and crises in various parts of the system. If those requirements are met, monetary transmission and payments flows will become smoother and more stable.

## ICTs and Securitization of Debt Instruments

The development of standardized and secure online messaging, and the substantial reduction of transaction costs and information asymmetry related to financial operations, gave a boost to financial innovation in various areas. In particular, that innovation is taking place in the transfer of risks from banks to capital markets through the development of new techniques for securitization of loans. That allows commercial banks to remove a part of the risks from their balance sheets and hence reduce the volume of capital that they have to maintain in order to meet regulatory requirements. The result is a thriving financial derivatives market and an overall increase in the securitization of traditional bank loans and other debt-related instruments.

While ICT-based innovation in finance was one of the drivers of financial globakzation and the increase in primarily private international financial flows, it further facilitated the process of liberalization and deregulation in financial services by blurring the frontiers between commercial and investment banking. More commercial parts of global banks started to resort to securitization of their loan portfolio, while investment banks and hedge funds were instrumental in marketing so-called credit default swaps and collateralized debt obligations and other debt-based securities, supporting the development of debt-related derivative instruments. The need to move a part of assets out of balance sheets to meet the stringent capital and asset quality requirements of Basel II has only accelerated that process.

Modern ICTs and, in particular, the Internet facilitated the securitization of traditional bank-related debt instruments and the dispersion of credit risks among non-bank financial institutions such as insurers, private equity and hedge funds. They thus permitted the development and pricing of a host of complex instruments and 24-hour trading of huge volumes of securities. The digitisation of financial operations and possibilities of creating hybrid products outside the scope of regulated banks triggered the creation of such instruments as collateralized debt obligations, credit derivatives, and structured investment products. In many cases these were managed by specialized companies and funds, special purpose vehicles and monoline insurance companies. While risk dispersion from banks due to financial innovation might be considered a positive trend per se, financial regulators are now concerned as to whether in the event of financial crisis the risk will flow back to the banking system and whether the latter will have enough capital and liquidity to absorb the stress related to such an eventuality In that regard overcoming information asymmetry at the level of global financial markets is important for the purpose of adequately assessing the distribution of risks throughout the international financial markets.

It is also thanks to modern ICTs that banks' trading activities related to various debt and other financial instruments became relatively more important than longer-term deposit-taking and credit-providing activities. While analysing the importance of the information revolution for financial innovations, especially in capital markets, is beyond the scope of this chapter and will, it is hoped, be tackled in forthcoming research, it was important to stress here the role of ICTs in facilitating securitized debt instruments, which are increasingly competing with commercial bank-based lending and borrowing operations.

## E-Finance and Security Challenges

A major challenge for e-banking and e-payments that requires further innovative approaches stems from the need to tame and contain the negative effects of rapidly growing cybercrime. Various actors involved in developing new security systems include technology vendors, security system designers, banks and non-bank financial service providers.

While according to an optimistic scenario about the use of e-banking in the United States the number of users will rise from 56 per cent of households in 2006 to 76 per cent in 2011, other sources are noting a tendency for some groups of customers to reduce or even cease their e-banking and e-payment activities because of security concerns. According to another source, 73 per cent of users in the 18-50 age group are concerned about identity theft.

The issue of security on the Internet has dominated more of the literature since the definition of e-banking moved from ATMs and telephones to the Internet. Security is all-important with online banking and it might determine its success or failure. In fact, some consumers stop paying bills and conducting other operations online as the (perceived) risk increases.

The security risk can cause banks to lose the gains of Internet banking if problems are not properly addressed. One of the risks is identity theft, which can occur in two ways: "phishing", namely sending aufhentic-looking-emails, which trick consumers into giving password details by replying to the e-mail; and "spoofing", which uses a fake website that makes customers believe they are using a real site. There are also key logging, software tracking and remembering passwords and numerous other viruses and trojans.

Another worry from the security point of view is the "trust gap", an unproven but suspected divide between those who use the Internet for other tasks and hence trust online banking and those that do not This means that, to encourage Internet banking, one may first need to encourage other activities on the Internet.

Once the banks start to outsource IT-related operations to other firms, there is the increased risk that the latter may not be sufficiently regulated and that the transfer of information to them and between them and the banks may increase the security risk. Consequently there is a need to develop approaches to make sure that a combination of enough technical expertise, security support and oversight is in place when one is engaging in outsourcing and offshoring in e-finance.

The security problem should, however, not be overestimated. Many innovative schemes have been developed to solve the problems, with more, including biometrics, in the pipeline. Furthermore, some authors claim that the slower growth of Internet banking in the United States is not linked to security concerns but to the macroeconomy and market saturation.

While ICTs help to improve credit risk management and hence the situation regarding information asymmetry, the technology itself can represent a risk and according to the BIS falls into the category of banks' operational risks. From the regulator's perspective, provisions to deal with the eventuality of service disruption for technological, including security, reasons are addressed in the operational risks cluster and are a prerogative of the Basel Committee on Banking Supervision. Maj or providers of payments systems such as the Society for Worldwide Interbank Financial Telecommunication (SWIFT), payment card providers such as Visa and Mastercard or peer-to-peer payments service providers such as Paypal are also taking major steps to introduce better-performing security systems so as to protect their clients. For example, Visa and Mastercard have improved security on the web with systems such as Verified by Visa and MasterCard SecureCode, while Paypal has made its security procedures more stringent.

Banks, other firms, institutions and Governments should adopt a risk management approach to information security. Its particular feature is that it involves an economic assessment of the information assets at risk, which is conducted before possible solutions are looked at. It should clearly outline security risks and compare these with the investment needed to guard against them, all the time considering the value of the underlying information assets. Policy processes at the national or international level may consider possible action to improve incentives for investing in better information security. More practically, in order to use a risk management approach, it is fundamental to define risks, to evolve ways of keeping risk perceptions current, and to measure or develop methodologies to quantify risks. It is immediately apparent that the task in question may be more difficult with regard to information security in financial services. Part of the problem lies with the ever-expanding scope of use of information technology in managing financial operations. Governments and financial industry supervisors and regulators may choose to mainstream certain aspects of financial information security risk measurement and methodologies, including the provision of quantitative data, in order to assist in policy development and implementation, as well as international cooperation on this crucial issue.

## Recent Trends in E-banking and E-payments

The traditional commercial bank services for corporate and household customers are undergoing dramatic changes that are moving rapidly towards e-banking. Given the relatively greater importance of commercial banking in the financial sector of developing countries, understanding the current trends in e-banking and e-payments, and deriving lessons from them, could be of the utmost importance for the further modernization of banking in those countries. Consequently there is a need to analyse the current trends and best practices in e-banking and e-payments, bearing in mind their applicability in developing and transition economies.

## The State of Play in Internet Banking

E-banking combines informational and transactional facilities. While the risks related to informational websites are limited to incorrect information or negative perceptions of the host bank, the risks with transactional facilities are more considerable. Those facilities should be functional and user-friendly, and customers should be sure that their security and privacy will not be infringed. In a wider perspective, e-banking risks include the following categories: transaction or operations risk, credit risk, funding and investment-related risks, legal and compliance risks, and strategic risk. Thus, the use of e-banking poses additional operational and hence reputation risk for existing banks, especially when they become vulnerable to fragile technology, customer confusion and hackers. Banks that start e-banking may also face initially high costs and technical problems, but those that wait, may lose customers to those who capture the market first. As the experience of many pioneers in e-banking shows, properly tackling the above-mentioned risks from the outset and managing them in the course of the further development of e-banking is important for maximizing the benefits of e-banking in the longer run.

The literature on e-banking mainly discusses the patterns of emerging virtual banks and of established banks diversifying from a purely branch-based approach to a twin online and more streamlined branch system that makes it possible to reduce high staff costs and overheads in a branch. The banks' strategies have also been reflecting consumers' preferences for a mix of delivery channels: the increasingly popular "click and mortar" approach. This model is becoming increasingly the norm in Europe and the United States, since consumers, although increasingly attracted by e-banking, continue to value personal interaction and are concerned about security risks on the Internet.

It combines face-to-face interaction and other advantages of traditional banking with the 24/7 availability and low, non-distance defined costs associated with Internet banking. Adelaar, Bouwman, and Steinfeld looked at a Dutch cooperative banking network and found that, since costs are much lower on the Internet, the banks had begun offering the most expensive services exclusively on the web. For other services customers can complete forms online and then finish the transaction in a branch. There is, however a risk that such "click and mortar" models will not be sustainable or will contribute to increasing the digital divide, since while educated and affluent customers will benefit from online services, the poor will lose out as fewer branches will be at their disposal. As a result in the early stages of e-banking there were calls for State protection of traditional banking to protect the poor.

It is now common knowledge that transaction costs related to Internet banking operations are much lower than those incurred at a branch or on the telephone. Many sources, including UNCTAD quoted the famous example showing that a payment operation that costs a branch one dollar could cost one cent if effected via the Internet. With these savings and relatively low set-up costs it is not hard to see the attraction of e-banking. Simpson used data from 1999 for a sample of American banks and banks in emerging markets to assess the different cost profiles and risks in e-banking and in traditional banking. He found that banks in the United States, which have reached an advanced stage in the use of e-banking have lower overhead costs and are considered to have a lower banking risk attached to them when compared with banks in emerging markets, which do not use e-banking Hence, banking risk and overhead costs could be reduced through greater use of e-banking.

The profitability of Internet-only banks has, however, been questioned, and statistical tests were used to compare it with that of established banks. In one model profit efficiency for Internet-only banks was higher, whereas in another it was lower, suggests that, at least in the short run, Internet-only banks are not necessarily viable and are more likely to be sustainable as one of the channels in a multi-channel banking services delivery model. In the case of Italy, studies have shown that there is a strong positive link between profitability and Internet banking activities. Internet banks may also have a revenue structure different from that of traditional banks, relying less on interest income and core deposits than other banks.

Internet banking cannot be forced on consumers, as many of them have the alternative of using their "brick and mortar" banks or telephones, and incentives are therefore needed. The factors influencing consumers to take up online banking vary between countries and depend on national culture, demographics and the structure of the economy. Li and Worthington examined data from 15 developed and 12 developing economies and found that the most important factor influencing the use of Internet banking is ownership of a personal computer: the more people in an economy who own a personal computer, the more Internet banking customers there will be. A lesser, but still significant correlation also exists between Internet banking use and the number of Internet hosts.

In developed economies the demand is for quality Internet banking provision, and this is reflected in its supply. For example, in the United States, Wells Fargo has developed a system called "web collaboration" which allows bank staff to see the same screen as the customer sees as they talk on the telephone. Wells Fargo has also introduced the "desktop deposit system" to allow online deposits without the need for extra software, and the "my spending report", which allows consumers to view simultaneously several accounts on one screen shot. In Switeerland, UBS, which is one of the largest global banks in terms of assets under management, has developed a sophisticated Internet banking menu for all types of customers, which allows them to make a variety of online transfers and payments, consult investment offerings and buy various assets. In parallel, in addition to ATMs, a system of so-called Multimat machines was installed in branches with screens enabling customers to undertake operations similar to Internet banking operations. As a result, the bank's branches are now staffed more with personnel who are becoming financial advisers who help clients to make more complex financial decisions such as obtaining credit or investing in various financial instruments proposed by the bank. Withdrawing funds or making payments is becoming an automated process. However, the e-banking systems face new problems - for example, clients sometimes have to wait to access to machines of which there are only a few, or customers, who are less knowledgeable about technology have to wait even longer to receive assistance.

In the area of risk management the use of technology means that the market risk incurred by banks (the risk that asset prices will change) can now be better identified and as a result of which a better risk-return trade-off is created. Customers also benefit from the banking system's greater resilience to risks. Banks are also able to evaluate and manage credit risk through the use of various instruments for debt securitization and the development of the derivatives markets.

### Wholesale E-payments

According to the Boston Consulting Group's regular global payments report, the volume and value of both domestic and cross-border payments in the Americas, Europe and the Asia-Pacific region will continue to experience dynamic growth. The number of domestic payments is expected to grow from 213 billion operations in 2003 to 414 in 2013, with an increase in value from $1,731 trillion to $3,146 trillion, while cross-border payments will display a similar growth pattern rising from nearly 2.5 billion transactions with a value of $318 trillion to 6.8 billion with a value of $489 trillion. Such volumes of payments traffic need to be underpinned by modern inter-bank electronic transmission technologies and techniques, based mainly on IP.

To ensure the fluidity of inter-bank payments, clearing and settlements systems of various levels are put in place, comprising mainly local and regional automated clearing systems that are backed up with larger RTGS systems mainly controlled by CBs. While banks domestically or regionally are interconnected through automated clearing houses (ACHs), at the global level such major inter-bank payment networks as SWIFT, Fedwire (US-centred Federal Reserve system), TARGET (Trans-European Automated Real-time Gross-settlement Express Transfer system) and some other RTGS systems represent the core of cross-border flows of payments. Messaging and transfers between financial institutions intermediated by those and other payment systems have been profoundly transformed by Internet-based technologies. In that respect, it is relevant to provide further information about the ability of CBs and commercial banks to keep up with the rapid technological changes and innovation triggered by the Internet revolution.

SWIFT, the largest global messaging system between banks (including those from developing and transition economies), started in 2002 its move from a "store and forward protocol" network to a "session initiation protocol" (SIP) network in the framework of SWIFTNet - the advanced IP based messaging solution of SWIFT. This was completed in 2004, and SWIFT is currently working with the Clearing House Payments Company (a leading system for settling and clearing globally United States dollar payments) to make the two networks interoperable, as the clearing house also upgrades to SIP. The new SIP network allows complete tracking of a transaction throughout the process. It also acts as an "actual private network", ensuring that messages are sent quickly and securely, with the component "FileAct" allowing bulk messages and cheque imaging. SWIFT supplied banks with specific pieces of software to enable them to transfer to the new system and in 2005 began fining banks, that had not yet migrated. Developing countries were slower to switch over because of SWIFT's technical requirements.

SWIFTNet is used by CBs including those of the United Kingdom, France and the European Central Bank. The Euro Banking Association (EBA) and several ACHs also use it, and it will be at the core of the Single Euro Payments Area (SEPA) when the latter becomes active in 2008. In addition to servicing financial service providers SWIFTNet is also used by major corporations to obtain information for their multiple bank accounts.

In January 2007 SWIFT took another step forward with its business model and launched a new access model called SCORE (Standardized Corporate Environment), which enabled, for the first time, corporations to interact in a standardized way with participating SWIFT financial institutions. Previously corporations could connect to SWIFT only through the MA-CUG (Member Administered Closed User Group) and Treasury Counterparty (TRCO) models, which continue to be available for corporations not eligible for SCORE. This new model is expected to cut down administration costs and simplify the process of standardized messaging as it essentially bypasses the banks when the corporation wants to send a message.

SWIFT's longer-term strategy is to increase the number of messages it carries and also the type of messages it can carry. The new SIP system allows several different types of messages to be transmitted. SWIFT has evolved to encompass securities with "Wall Street related" messages, which account for almost half of its network traffic. This longer-term strategy is focused on securities and derivatives, once SWIFT feels it can add value through standardization and hence challenge Euroclear's domination.

**Retail E-payments**

Regarding the retail e-payments, the literature reported inter alia the "non event" of the move to new methods e-money. In developed countries most e-money models have failed while traditional payment methods, including credit transfers, cheques, direct debits and payment cards, have evolved into their electronic forms, this evolution entailing their active migration to the Internet. Nevertheless, credit cards, giro and in some countries cheques are the main methods used in retail payments. New integrated or cumulative payment services such as Paypal and new payment portals for consumers and small businesses have also been introduced. Moreover, mobile telephony is increasingly becoming a preferred platform for m-banking and inpayments, which may use payments messaging systems that are in many cases compatible with the Internet.Online credit transfers and giro, bills and cheque payments are the main Internet-banking services provided to retail customers. Another service that banks provide on a recurrent basis is direct debits, namely payments by banks to various service providers according to the customer's instructions. Standing orders are similar to direct debit operations and are initiated by the payer rather than the bank. Electronic bill presentment and payments (EBPP) services offered by financial and other institutions are also emerging as a major method of making and tracking online payments.

The global payments card market continues to be dominated by a few players, such as Visa, MasterCard, American Express and Diners. The legal decision handed down as the result of the challenge of American Express concerning MasterCards and Visa's practices, especially by setting of high interchange fees, was supposed to lead to increased competition in the market. However, so far, those fees continue to show rise. According to the Nilson Report, a consultancy specializing in consumer payment systems research, the fees that the United States merchants are paying to card companies and their issuing banks have been constantly increasing during the present decade. Thus, the weighted average fee rose from 1.52 per cent in 2000 to 1.88 per cent in 2006, while the volume of fees collected for the same years jumped from $24 billion to $56 billion. As a result, the merchants filed a class action against major card companies. In spite of the increased competition from various debit card networks, automated clearing house networks and systems such as Paypal, it is still the issuing banks that have the bargaining power when deciding which card to issue. As a result, the major part of interchange fees received by payment card companies goes to the issuing banks.

The Initial Public Offering (IPO) of Mastercard is not only providing the means to pay the legal charges related to the above process, but is also a part of a change in MasterCard's business strategy to a more open form of business and a new corporate governance structure intended to meet competition in the credit card market. After the IPO it changed its official name to "MasterCard Worldwide" and now has a three-tiered business model as a franchiser, processor and adviser. American Express and Discover take a more segmented approach, looking at the upper and lower ends of the market respectively. These three players are all public companies, while Visa like SWIFT is still a cooperative of banks servicing over 20,000 financial institutions. At the end of 2006, Visa, the largest global payments card company, announced restructuring whereby Visa Canada, Visa USA and Visa International would be merged to create a global public corporation called Visa Inc., the eventual goal being to launch IPOs. Meantime Visa Europe will probably continue to remain a membership organization. In consequence, Visa expected to accelerate product development and innovation.

Although they are convenient means of payment, credit cards continue to be expensive not only for merchants but also for those cardholders that use them as a short-term consumer credit instrument. The interest on credit card overdraft is at least 15 per cent per annum including in countries where interest rates, mortgage rates included, are much lower. In spite of the unsecured nature of card-related debt, such a large difference is not justified, especially in view of the new possibilities arising from current small borrowers' credit-scoring techniques. In fact, one of the reasons for mortgage-related debt increase in the United States, and one of the underlying causes of its current crisis, is the massive refinancing by US households of their consumer debts (i.e. mainly credit-card-related debts) through incremental borrowing on the mortgage side using the collateral of the house.

To counter this lock-in effect of credit card market oligopoly, competing debit card systems, frequently initiated by major retailers, are emerging and new card issuers are entering the market. However, they are far from being the leaders in acquiring a critical mass of clients, and hence convincing the banks to issue the cards on their behalf. Moreover, the major credit card companies and major issuers are also entering into ventures of presenting major retailers, and in such cases they agree not to charge annual fees and considerably reduce the interest rates charged in the case of overdraft.

Despite the problems outlined above, the convenience of electronic payments proved to be a stronger argument for consumers, and in the United States for example, in 2003 the number of electronic transactions (card and direct debit transactions) totalled 44.5 billion, and overtook for the first time the number of cheques paid, which stood at 36.7 billion.

Lastly, mobile hansdsets are increasingly becoming a means of payments for their owners – either through their use as a credit card, with the customer being billed by a mobile operator at the end of the month, or as a prepaid card. Prepaid phones, by definition, use the latter form of m-payment. The active involvement of mobile network operators in such payments has created some legal uncertainty as far as their role in the financial sector is concerned, and regulators are trying to address this question without impeding innovation in that sector. In fact, competition from non-banks in payments services is exactly what

## E-Banking and E-payments for Development

### E-banking and E-payments in Developing and Transition Economies

Since banks in developing countries and in countries with economies in transition did not have an abundance of traditional proprietary technologies and hence less attached to particular banking techniques, they were also in many cases more open to introducing the most recent Internet-based technologies. The fact that they were less burdened with tradition in some cases made it possible introduce new technologies more rapidly, and also adapt to them more rapidly. However, that requires more vigorous training and retraining of employees. This is why it is important to understand how the financial institutions in those countries were managing to strike a balance between the opportunity to be more innovative and the challenge to service a population with a lower per capita income and large share of the "unbanked". In particular, it is interesting to identify how using ICTs and providing e-banking and e-payment services could improve the access to finance of SMEs and microenterprises, which represent the core of those economies. This underlines the importance of further research into the use of e-banking and e-payments in SME finance, as well as in micronnance.

There are a number of features that the customer values in a financial service. These include acceptability, accessibility, affordability and ease of use. As far as businesses are concerned they look for a number of other features, ones that will help ensure that they provide a cost-effective service. They include functionality, segmentation (the ability to sell different products to different groups of customers), competitive fees and charges, increasing efficiency, controlled development costs, distribution among the population and partnerships with other companies. Financial literacy is important for increasing e-banking among the poor, and there are a number international programmes, for example set up by the World Bank, to increase this literacy.

Much of the literature on e-banking focuses on case studies and comparisons in developed countries, where it is well established and in some cases is not far form the point close to saturation. However, e-banking is also increasingly taking root in many developing and transition economies, and in some of them its penetration exceeds OECD median indicators.

For example, Internet banking has been surprisingly successful in Estonia. The first bank to introduce Internet banking was Hansabank in 1993. There are several reasons for this success. First, since the country had previously had a command economy, the traditional commercial banking did not exist, and hence consumers did not have time to grow attached to the branch model used for many years in other countries. As a result, consumers did not value or expect much from the bank branch experience. Second, banks were still developing and did not have full country coverage. They lowered their costs and improved their image. The success of Internet banks has been largely due to their wide appeal and the Internet technology available for reaching sparsely populated areas; that availability has been due in turn to government policies to deregulate, increase competition in the telecom markets and increase Internet use. In fact Estonia was ranked quite high - ahead of some OECD countries - in the Networked Readiness Index of the World Economic Forum and the E-Readiness Ranking of the Economist Intelligence Unit.

Internet banking has grown rapidly in Braxil and has been the fastest-growing banking medium. In Turkey e-commerce is also most developed in the banking sector and driven by reduced costs, transaction costs on the Internet being only 5 per cent of those at a branch. The first bank to use the Internet in Turkey was Isbank in 1997.

The Nigerian Inter-Bank Settlement System (NIBSS) has recendy developed the Electronic Fund Transfer Service, which focuses on giro and ACH functionalities. NIBSS is collectively owned by Nigerian banks and is responsible for transactions including final settlement with the Central Bank. Launching ACH and adopting the electronic cheque presentment with image exchange system have shortened the clearing cycle still further, as well as reducing fraud and cutting costs. Despite great ambitions for Internet banking in Nigeria, these were thwarted by technical problems. This was particularly the case with the national telecommunication network, when banks have been led to seek alternatives to it, but in an uncoordinated and hence cosdy manner. Nevertheless, consumer perception of the effect of IT on the banking industry stayed positive.Thanks to the policies of the Central Bank and the well-developed business sectors, South Africa is particularly well suited to electronic banking, which includes ATMs and other features. Those features make them attractive to poorer customers. However, there is also the psychological barrier, especially among the poorest. Singh and Malhotra conducted a survey to determine why some South African customers are reluctant to bank online. They found that the more affluent and younger are the groups that are most frequendy banking online. For those groups that did not bank online, the main reasons for not doing so were fears about security and lack of knowledge. At the end of 2003 online bank accounts in South Africa topped the million mark for the first time, amid claims that the media had grossly exaggerated security fears but been mainly ignored by the public.

The Reserve Bank of India (which is the Central Bank) has been heavily involved in the development of e-banking and has drawn up guidelines for banks to follow when providing Internet banking services. The development of Internet banking in India began with the foreign and private banks, with the public banks lagging behind.

In Romania foreign banks also led the way in electronic banking, beginning in 1996, and local banks followed later; however, usage rates have remained low. Banks need the approval of the Ministry for Information Technology and Communication before they can offer Internet banking services. There are currendy 28 banks offering such services. Most of the approved banks use VeriSign to guarantee their sites and also employ methods such as Digipass Pack to identify users and secure transactions.

Siriluck and Speece suggest that Asian cultures attach greater importance to interpersonal contacts and relations than those in the West, and as a result Internet banking faces an extra hurdle. Malaysia is especially interesting for Internet banking because of the high level of technical education and the quality of the technology available. In a survey of over 500 people in that country the use of online banking was seen to be concentrated around younger and higher-income groups. The quality of online banking websites was high, although not all banks provided all services and options via their websites. Most Internet customers had a positive perception of Internet banking; however, there were also security concerns. Foreign banks were allowed to provide banking over the Internet as from January 2001 (a year after domestic banks). But this delay did not protect domestic banks from foreign competition, even though foreign banks offered less technologically sophisticated services). Regulation took the form of the Central Bank's minimum guidelines on the provision of Internet banking issued in 2000, which require banks to meet with customers face to face before an Internet account can be opened or credit given. The Central Bank of Malaysia has developed research collaboration between Asian CBs and the BIS on technology issues, and has also established minimum guidelines and standards for risk management as well as monitoring of compliance and risks.

Electronic cards are becoming one of the channels for e-payments in countries with a low level of banked population. Figure 1 shows transaction flows related to regional payment cards project of the West African Central Bank (WACB) created within the framework of the Economic and Monetary Union of West Africa. The common currency in the Union helps banks in the region to avoid foreign exchange risk in their mutual transactions and has enabled WACB to develop a card-based payment system called "Monetique UEMOA" (in addition to other two regional RTGS and ACH systems). The system has an inter-bank card payments processing centre that allows 63 commercial banks in the region to issue payment cards for corporate and individual users.

The phenomenal growth of mobile telephony in developing countries has also brought about a dynamic expansion of so-called m-banking and m-payments. They are becoming very popular in small purchases through m-commerce as well as in P2P transfers such as remittances, or in B2B payments, especially in microfinance. Since most mobile phone owners in developing countries are "unbanked", they participate in e-payments by buying prepaid cards at various points of sale and then make their m-payments. Given the relatively high level of mobiles' penetration in developing countries, they can become the main channel for delivering online payments, especially in countries and regions with few branches and ATM networks.

A mobile payment scheme called M-PESA is being successfully run in Kenya by Vodafone, local Safaricom with the support of the United Kingdom Department for International Development. Around 150,000 M-PESA customers with the help of around 500 agents, can make payments or send money using their mobile phones. The customers' phones are charged with prepaid funds in Safaricom shops, petrol stations and other points of sale as soon as the owners of the mobile phones, whose identities are verified via their mobile numbers, pay the cash. M-PESA has developed its own financial system, including clearing and settlement, and keeps the float in a single account in the Commercial Bank of Nairobi. As long as it does not pay interest or invest this money, the Central Bank of Kenya tolerates such financial activities.

The following sections will focus on three important aspects (from the development perspectives) of financing destined to less endowed groups of population and enterprises and namely remittances as well as e-finance for SMEs and microenteprises. The introduction of digital delivery channels in those areas is providing new opportunities to combat poverty by improving access to finance through the better use of new more accessible for those groups tools of ICTs.

## ICTs and Remittances

Remittances, representing mainly small-scale transfers by migrants, expatriates and to a lesser extent charities, are becoming an increasingly important source of external financing for many developing and transition economies and as such merit careful examination. As a distinct item of balance-of-payments (BOP) financing their volumes in the case of many recipient countries are comparable to other sources of foreign exchange such as exports, foreign private capital inflows and official aid. However, unlike others, those flows, also known as unrequited or unilateral transfers (i.e. not incurring debt-related obligations), are a result neither of earnings by country residents due to exports of goods and services, nor of their borrowing or investment-attracting activities.

According to the World Bank, remittances sent by migrants to developing countries reached $206 billion in 2006. However, it is estimated that the real level of current remittances could be 50 per cent higher owing to the major role of various informal remittances networks. In other words, the overall volume of all remittances might come close to that of FDI. According to UNCTAD, FDI flows to developing economies totalled $379 billion.

Compared with other types of financial flows, remit-tances are the least cyclical source of BOP financing. They play a stabilizing and even countervailing role when countries, in adverse economic circumstances, suffer from reductions in flows of external finance coming from other sources and especially from short-term portfolio capital. That positive effect is due to the underlying motivation behind those transfers, which is primarily to support families back home. Remit-tances allow the consumption level of the poor to be sustained, giving them better access to basic facilities, including health and education facilities, in recipient countries. Remittances also help to launch and sustain small family businesses.

Historically the main part of those funds was transferred through informal systems such as hawala. Hawala is a form of informal networking whereby hawala brokers intermediate transfers without actually moving funds. They work on the basis of trust with minimal documentation, and each tries to have enough funds to meet the demand of correspondent brokers). As in the case of credits, informal intermediaries were charging the remitters passing through their channels high fees. At the same time, the fees collected by major formal money transfer service providers were also initially quite high, considerably exceeding, for example, the level of interchange fees paid by merchants for credit-card processing. However, with the advent of the Internet and further dissemination of electronic means of payments, various money transfer service providers started to carve large chunks of payments out of the cash-based systems. Competition, and in particular the use of the Internet, made it possible to considerably reduce fees charged by such major operators as Western Union and Moneygram. The competitors from regions within the main corridors of remittances were quite creative in using various delivery channels, including Internet-based networks or m-payments, and hence considerably cut their costs of operations. They were also less strict than major Western operators in their perceptions of risks, given the extremely short periods of time needed to complete the e-remittance operations.

Regional systems such as Anelik and Unistream were charging much less for transfers from the Russian Federation to Armenia than was the case for Western Union. That competition has paid off and while for the same corridor Anelik is still charging around 3 per cent for the transfer of, for example, $200, Western Union cut its fee to around 5 per cent, which is much less than in 2004.

In countries with large migrant populations banks are trying to attract migrants as bank clients and are using remittances as a tool to attract more deposits from them. For example Chase has introduced a scheme for the free transfer of money to Mexico, enabling cheque-account customers to make a limited free transfer of funds (up to $1,500 per month) to its partner bank Banorte, which has 1,000 branches and 2,800 ATMs in Mexico. The recipients of those funds can either keep them in accounts in Banorte or collect cash without paying a fee. According to the Central Bank of Mexico, remittances from the United States totalled $23 billion in 2006 and are expected to reach $24,5 billion in 2007.

A relatively new phenomenon is the use of mobile phones as a means of transferring money between countries or within developing countries from urban to rural areas. Sometimes those areas may even lack bank branches, and the only way to cash the transfer may be to use the specialized points of sale that also sell prepaid cards.

**E-banking and e-payments for SMEs**

To access finance, SMEs have to compete with households and large corporate businesses in order to gain a share of the financing disbursed by banks. Traditionally, banks were biased against SMEs given the relatively higher costs of acquiring and processing transactions for SMEs in the light of a wide geographical spread and the greater number of low-volume transactions. That resulted in high unit transaction costs related to SMEs' financing. Modern e-banking with online credit risk databases and much lower transaction costs related to data processing and data mining is providing new opportunities for improving SMEs access to, in particular, short-term working capital and trade finance through e-banking and other e-trade finance services. It is becoming possible for SMEs to manage their own accounts and payments traffic online, while banks can not only provide finance, but also help them manage their cash flow and ensure a quicker turnaround of working capital.

For financial institutions, servicing this sector requires the finding of solutions that take into account the specific risk-return perspective of the sector. For the most part, the challenges to SME funding result from lack of information, such as unclear financials, and the fact that information about SME borrowers is being limited. There is also a lack of comparative data on SMEs. Banks in most developing countries do not have access to transaction data and credit history (unlike in many developed countries) through credit bureaux and other credit management service providers such as credit insurers and factors. Developing such services primarily through the appropriate use of ICTs - offering financing and credit management through open networks such as the Internet - could overcome impediments related to SME financing.

E-banking benefits are especially apparent to banks when doing business with SMEs. SME business was labour-intensive and offered relatively low profits for banks, whereas SMEs were finding the service provided by banks too "one size fits all", inefficient and supply-driven. E-banking and other e-finance techniques might address those problems through automated, 24-hour procedures and lower costs. SMEs can then use their new efficiency and access to credit at lower rates to fuel growth in developing countries. However, as Wendel and others emphasize this debate is not much different from the debate related to retail banking. The role of the traditional branch, especially when dealing with SMEs suggest that the "click and mortar" approach will continue.

In addition, the reduction in costs associated with the possibility of having better-organized information flows, greater transparency and hence a better loan-tracking process can be considered to be positive externalities of ICT adoption in SME financing. The new e-fmancial tools that are designed for SMEs and their needs are supposed to help in gaining those advantages. Even though the development of those systems is not very time-consuming, its finalization requires quite a long time, and the systems cannot start to be remunerative until their integration process has been completely finished.

Investments in this field cannot be remunerative unless they are associated with a change in business process and improvements in skills at different levels. As a consequence, ICT use can be considered a new approach for both human capital and the organization of SMEs, and the means to achieve better financial results. Local requirements are certainly one of the most important issues for the success of this technology in developing countries. Delivery channels and product mix are two important aspects for the success of each company. However, the high costs of the comprehensive adoption of ICTs often imply the need for public financial support or various forms of public-private partnerships with development and export-import banks and other supporting institutions.

In order to create a public record of past payments SMEs in developing countries need to move out of informal economy. By fully formalizing themselves SMEs might create for themselves a public reputation as of good-quality borrowers. Under such circumstances financial institutions can create electronic credit histories and Internet-based credit risk databases in helping creditors to provide capital with fewer and less stringent requirements. A responsible behaviour by firms can lead banks to trust SMEs more, requiring fewer guarantees for loans and creating conducive borrowing environment. As a consequence, such a change in SMEs way of operating can have important effects for the whole banking sector. Indeed, there is a need for a larger formal economy, characterized by a business-friendly regulatory and institutional environment, including a more competitive banking sector that will provide credits to SMEs with less onerous collateral requirements and with lower interest rates. A larger formal market, including the majority of SMEs, would also attract other financial service providers that will contribute to credit risk management and financing; this will lead to more competition in the financial sector.

The lending infrastructure plays a very important role in the efficiency of ICT use. Indeed, the introduction of internationally acceptable accounting and payment performance information of a good standard are fundamental for credible financial statements and are a precondition for the feasibility of loan contracts. Also, the legal and institutional infrastructure strongly influences the adoption of ICT necessary for backing up e-finance for SMEs. Commercial codes, property rights and bankruptcy laws affect the degree of confidence in financial contracts. Moreover, even tax and regulatory environments can alter SME credit avilability in a negative way. In some cases, regulatory restrictions on the entry of foreign banks into the domestic market and other restrictions on foreign banks' operations may negatively affect SMEs' access to credit. As a consequence, the availability of credit for SMEs depends on the quality of regulatory and institutional environment in general and lending imitations in particular.

The structure of financial institutions impacts on SME lending and can lead to different treatment for SMEs operating in similar markets. Large institutions can run economies of scale and offer competitive interest rates, even though they often analyse firms only from the financial ratios perspective. However, large institutions certainly have more agency problems than smaller ones characterized by fewer internal hierarchical separations. As a consequence, the latter can manage soft information more easily than larger institutions. In developing countries it seems consistent for foreign-owned institutions to have more problems than domestic ones in managing soft information. Nevertheless, a combination of better technology, easier capital market access, greater capacity to diversify portfolios and better working skills can make foreign-owned banks more efficient. Also State-owned banks in developing countries, in order to succeed in financing SMEs, need to pay more attention to modernizing their delivery channels through better use of ICTs. The use of various modern ICT-intensive delivery channels and more competition might attract new lenders in the SME finance market and result in better access for SMEs to e-banking and other e-finance facilities.

### Microfinance

Micronnance, by definition, constitutes credit and other financial services of a low monetary value given to microenterprises and households primarily to encourage their productive activities as a means overcoming the poverty trap. In addition, the aim of microfinance institutions (MFIs) is to reach out to the poor, who have been left outside the traditional mainstream financial services. This implies that a very high volume of small transactions is a defining characteristic of MFI operations.

As in the case of SMEs, the main reasons why commercial banks do not regard the population that microfinance targets as worthwhile customers is the fact that serving this population involves a high volume of low-valued transactions. Traditionally high transaction costs and other market failures were a barrier to supplying small-scale credits, even though the interest rates charged could be relatively high. However, the cost of formal microfinance credits are still normally much lower than the usurious terms charged by informal creditors, especially in rural areas. In fact, the main target populations for microfinance are peasants, and tapping into that market is a strategic goal for microfinance.

One of the reasons for the cost of microfinance that is acceptable to borrowers is the fact that many MFIs receive funds from NGOs or donor agencies. However, that is starting to change. While many are still donor- or public-investor-dependent, others are becoming more market-oriented, profitable and (it is to be hoped) sustainable. The proliferation of various types of MFIs is bringing about a rapid increase in their numbers in developing countries.

One of the key factors distinguishing top-level MFIs from their peers is the adoption of technology. In fact, some MFIs are becoming quite sophisticated in their use of ICTs. The obvious benefit that the MFI industry derives from ICTs is the reduction in costs terms of the MFIs' back-office operations and of delivering their services, which should mean lower fees for MFFs customers. However, there are additional benefits derived from the introduction of ICTs, such as easier tracking of loans and repayments, standardized processes and better flows of information in organizations, and more transparency. Also, MFIs may be able to offer additional features and services that would not have been possible without the adoption of ICTs.

With regard to clients, ICTs can be used to expand the range of products and services offered by MFIs to their clients. The main technologies that have been identified are ATMs, point-of-sale (POS) devices and mobile phones.

There are different types of ATMs: they range from those that dispense only cash to those that offer a full range of services, such as accepting deposits and transferring money to different accounts. ATM networks are generally quite expensive to own and operate, al-though costs are falling thanks to improvements in technology, which may no longer be as prohibitively expensive as before. An ATM network allows customers to conduct transactions at their own convenience, and provides banks with a cheaper way of handling a high volume of transactions.

A POS network involves devices installed in stores and other appropriate oudets which enable customers to make their payments and transactions using their cards. These cards can be of the traditional magnetic-strip variety, or the more advanced, and more expensive, "smart cards" with an embedded chip, which can contain detailed transaction records and personal data to enhance security and facilitate transactions. Cards can also store value on a prepaid basis and be used as a means of payment. POS devices tend to be cheaper than ATMs, and the outreach of a network of POS devices installed in small stores and shops can be greater than the outreach that can be achieved through an ATM network. As in the case of ATMs, such a network allows the MFIs to process a large volume of transactions more cost-effectively than over-the-counter operations. However, a POS network tends to be limited regarding the type of transactions that can be handled, and there are operational, and even regulatory, complexities that arise from the fact that retail oudets are in effect acting as agents for the MFIs.

Lastly, mobile banking is an especially attractive channel in countries where a fixed-line network is not well established. In much of the developing world, mobile telephony has overtaken landlines in terms of access, volume of calls, and even affordability and reliability. People that own mobile phones but who have never had a bank account are now a widespread phenomenon. Some MFIs have tapped into this opportunity so as to offer microcredit loans and other financial services via menu-driven systems or via SMS text messaging The advantages of these channels include simplicity and user-friendliness, and the fact that the infrastructure is in many cases already in place.

Apart from lowering transaction costs, the primary additional benefits of having these new channels of delivery are the increase in the volume of transactions by increasing their accessibility and expanding outreach, especially to more remote places, and, by extension, the increase in the convenience for customers.

As in e-banking in general, a number of factors have been identified that determine the value attached by customers to moving away from cash to electronic solutions in microfinance. The additional features offered by the new technology to customers include acceptability, accessibility, affordability and ease of use. Stemming from these features, the key drivers in determining the success or failure of client adoption of a new technology within the context of microfinance are its perceived increase in value; the level of education and training of the customers, and thus their comfort level in using this technology; its usability; the cultural fit of the technology within the community; and whether or not clients have sufficient trust in the new technology.

To reach greater levels of efficiency in organizations, ICTs should be integrated into the management information system (MIS) to make it possible to automate the flow of information and thus ensure the smooth running of agreed procedures and networking between employees. Although this is technically incorrect, the MIS of a bank or financial institution is treated like its back-office operations. For an MFI, there are three key areas of an MIS: the accounting system, the credit and savings-monitoring system, and the client impact data-gathering system.

Existing software packages for the accounting system are abundant and easily found. They differ widely depending on the specific needs as well as legal requirements of the MFI and the country in which it operates. The decision on a credit and savings monitoring system, however, is often more complicated. There are no real standards or guidelines. This is because a well-thought-out system should fit the needs and operations of the MFI using it. MFIs differ from one to another and hence there is no off-the-shelf software that can address the requirements of every MFI. They are at the same time less complex in terms of operations, than commercial banks (Ahmad, 2006). The software should facilitate transfer of data to and reconciliation with the accounting system. Integrated systems do precisely that, but they are usually quite expensive, especially if a high level of customization is required. The third system, for client impact, is usually designed informally, if at all (CGAP, 1998, 2001).

In order to maximize the benefit of cost efficiency and for the technologies to be sustainable, there needs to be a sufficiently large volume of participants. This calls for a shared infrastructure among MFIs. However, experience has shown that sometimes MFIs resist cooperation for fear that their competitive advantage will be lost (Firpo, 2006).

There are benefits in addition to increased efficiency and lower operating costs that MFIs can reap from technology - for example more informed decisionmaking, increased flexibility, and better transparency and reporting, the latter being especially important for MFIs that are dependent on donor funds (CGAP, 2006).

Generally, MIS can be categorized into three types: manual systems that are still in being used by very small organizations; semi-automated systems, which do not completely satisfy organizations' information requirements; and fully automated systems. Most MFIs are in a semi-automated mode. No matter what the level of automation is, for an MIS to be effective, it must be cost-effective and fulfil business requirements, while being flexible regarding future changes, reliable, simple to use, scalable to accommodate the growth of the business, and integrated amongst the branches of the business to produce a single consolidated picture (Ahmad, 2006).

Because many MFIs have branch offices, as part of their mandate to reach out to geographically remote areas, the issue of networking needs to be addressed. The World Bank suggests that each branch should at the very least have its own MIS and database. If resources allow, the MFI should consider networking all branches together, which would allow real-time access to up-to-date information, which in turn would greatly improve information efficiency. However, setting up and maintaining a network can be quite costly, adding another level of complexity to its day-to-day operation. Moreover, many of the branches in the rural areas of the developing world are constrained by the level of reliable communication infrastructure available, and establishing a network may not be feasible at all. Furthermore, in locations with harsh environments, the need to support and maintain hardware equipment in the branches may even outweigh any benefits provided by the technology.

The recurring lesson from the experience in the field has been that ICTs are not a fix-all solution. A careful analysis of an MFIs' operations and its exact ICT needs should be conducted before an investment decision on the implementation of a new system, be it on the client delivery side or in the back-office. So far, even among leading MFIs, using technology productively is still an exception rather than the rule. Therefore, while it is recognized that most MFIs do eventually need an MIS system, it may not always be abundantly clear to decision makers that such an investment is needed. A balance needs to be sought between solutions that are appropriate and those that are state-of-the-art but are of limited practical use because of constraints in the local context. Such constraints could be anything from an inadequate telecommunications infrastructure to the prevalence of illiteracy.

Since in general the cost associated with building the infrastructure for the new technology is too high for MFIs to bear on their own, external funding sources are often needed. The management of the MFI and the donor organization or outside investor need to ask questions about the proper fit of any proposed adoption of technology with the strategy and structure of the institution, as well as carefully weighing the monetary cost and the ongoing resource requirements needed to maintain the system. As part of that process, Ivatury makes four recommendations about how to overcome the common challenges that an MFI faces in tackling technology projects. First, manual processes should be improved and streamlined; second, the decision makers need to ask the right questions; third, independent expert advice should also be sought; fourth, there should be plans to train staff on an ongoing basis.

## Regulatory Issues

E-banking and e-payments are increasingly being scrutinized worldwide by regulators that are trying to ensure financial stability without impeding technological innovation, especially in the banking industry. The main CBs are coordinating their efforts in this field with support of BIS through the Basel Committee on Banking Supervision (the latter deals with IT-related operational risks), the Committee on Payment and Settlement Systems and other forums. Consensus achieved here is not binding for developing and transition economies' CBs, but the latter normally consider above frameworks as those generating best practices and normally follow their recommendations. The CB's have a dual role as facilitator and regulator of e-banking. The United States Federal Reserve is an important example, as it operates an ACH and Fedwire. It has introduced FedLine for the web and other electronic products. Regulation is especially important now since e-banking may be subject to security breaches that can undermine confidence and thus the stability of the banking sector. In terms of regulations the Basel Committee has laid down guidelines, which the CBs have elaborated on. For example, the Federal Financial Institutions Examination Council in the United States has developed them in more detail and the Singapore Monetary Authority has developed its policy statement on Internet banking. But more legislation is needed, including on such essential elements as electronic signatures, money laundering and others.

There are several tools that regulators can use. These include existing regulations and laws adapted to the requirements of above-mentioned Committees, and making sure that banks have adequate facilities for training staff in new technologies. The legal definition and recognition of new methods of conducting transactions and the international harmonization of regulations are important for preventing a "race to the bottom" and also properly defining the roles of host and home country jurisdictions as regards international banks and cross-border customers. It is important that new risks and challenges be integrated into the evaluation criteria for banks. Ironically, regulation is most needed where it is hardest to implement, as E2eoha explains in the case of Nigeria, where there is a need to protect an emerging banking industry and also the country's international reputation by addressing "Internet fraud" and deeper problems in the politico-economic system.

However, even once laws are in place careful implementation is needed. In South Africa banks have been accused of unjustifiably requesting online customers in cases of identity theft to sign away their rights to money that might be eventually returned. Also, the standards of banks' security training and advice on their websites are still relatively low. Another regulation-related issue is the need for a comprehensive system. For example, in India the data protection law emphasizes that hackers should be punished. However, firms are under no obligation to prevent them from being able to hack into the systems in the first place. Thus the law does not really serve the consumers' interest. Here also legislation prevents anyone other than a security guard and the customer from being present at an ATM, and thus new customers cannot be taught to use ATMs. In spite of major advances in financial sector liberalization, Indian banks were not allowed to compete with rural regional banks, and hence expand a network of ATMs to optimize their value.

In South Africa security is still a major problem, especially since organized crime groups have developed software to steal passwords and identities. The law is uncertain, both in its terminology and whether or not it applies to online banking. South African banks have experimented with several measures. Nedbank has an SMS aufhorization system whereby a payment needs to be secured by the customer's providing a unique reference number sent to his or her mobile phone before it can be completed. This is mobile-phone-based and hence avoids viruses and data miners on the customer's computer. Standard Bank, the second largest bank by market share, has also developed a similar system, whereby passwords are supplied to phones and not via a computer.

In relation to self-regulation, the World Bank has made a number of policy recommendations for e-banking These include establishing a comprehensive security control process, centralizing back office operations; developing an automated credit aufhorization system by developing an appropriate credit scoring system; comprehensive oversight of outsourcing partners; and proper inclusion of e-banking risks in the overall risk assessment. Authorities should provide oversight of security controls and of outsourcing and partnership arrangements, provide the infrastructure and human capacity for technology adoption, and create appropriate regulatory frameworks. By doing that they strengthen the payments and settlement systems, as well as improve transaction-reporting services. Also, they should use new technology to better disseminate information, especially relating to credit, and to establish set up registers for collateral. Regulation by the public sector is needed, they argue, in order to organize security procedures and standards for electronic signatures.

As regards at the "bigger picture", the role of CBs has also changed with technology. They need to adapt monetary-policy-setting rules related to the use of new technologies. Since e-banking makes transactions cheaper and easier, users of money do not have to hold cash balances for as long, and incur the losses of doing so. There is also increased access to interest-bearing assets via the Internet, especially to

households and SMEs, and thus there is more incentive not to hold cash balances. This means that money demand is more sensitive to interest rate changes. The role of the CBs is to ensure, together with monetary stability, the stability of the banking sector. This means they must adapt to the changes in banking technologies and techniques. That includes developing secure and technologically advanced payments and setdement procedures, especially cross-border ones. Broadening the scope of CB research beyond the effect of their policy on the real economy is one of the important steps in that respect. The continued implementation of the Basel accords should assist in this transition. Finally, there is a need to recognize the bypassing of banks by non-bank electronic payments providers, and thus a need to develop a regulatory oversight over those institutions as well.

## References


Ahmad A (2006). *Management information systems (MIS) for microfinance.* Foundation for Development Cooperation, Brisbane, Australia.

Arumuga S (2006). Effective method of security measures in virtual banking. *Journal of Internet Banking and Commerce* 11,1.

BIS (2004). *Survey of Developments in Electronic Money and Internet and Mobile Payments.* Committee on Payment and Settlement Systems, Bank for International Setdements, March.

GAP (Consultative Group to Assist the Poor) (1998). *Management Information Systems for Microfinance Institutions: A Handbook.* Technical Tool Series No.l, CGAP/ World Bank.

Cracknell D (2004). Electronic banking for the poor: panacea, potential and pitfalls. *Small Enterprise Development,* vol. 15, no. 4, pp. 8-24.

Firpo J (2006). Banking the unbanked: technology's role in delivering accessible financial services to the poor. Mcrodevelopment Finance Team. Development Gateway Foundation, Washington.

Furst K, Lang WW and Nolle DE (2002). Internet banking: developments and prospects. *Journal of Financial Services Research,* vol. 22, no. 1-2, pp. 95-117.

# 10

# Role of Telecentres in Information Economy

The multi-purpose applications of ICT and the rapid fall in its costs have allowed innovative uses of that technology in many poverty reduction programmes. Governments worldwide are developing and implementing ICT policies to support economic and social development. However, so far, the impact of ICT strategies and programmes in providing economic opportunities for people living in poverty has been limited. Attempts to provide wider access to ICT, including the establishment of public ICT access spaces (telecentres) and programmes to develop ICT skills, have not been sufficient.

This chapter reviews the role of a key policy instrument (telecentres) in promoting livelihood opportunities for people living in poverty. To examine how ICTs can support means of living for people living in poverty, this chapter will use the concept of livelihoods. "A livelihood comprises the capabilities, assets (including both material and social resources) and activities required for a means of living". In other words, what is important for a person is not only the wage income that he or she may receive but also all the other assets (e.g. subsistence food, savings, physical assets, access to government support) and capabilities (e.g. education, access to markets, access to information, ability to communicate) that he or she may need to have in order to earn a living.

Sustainable livelihoods approaches put people living in poverty at the centre of development programmes and emphasize that programmes supporting poverty reduction should be based on the specific context, and people's vulnerabilities, capabilities and assets, and should address the influence that institutions (whether Governments, civil society or private companies), organizations and social relations have on the ability of men and women to pursue a strategy to escape from poverty. Thus, using telecentres for promoting livelihoods requires an understanding of how they can support people's capabilities and assets, how they can address their vulnerabilities, and how they can work in and influence the specific institutional and social context shaping those livelihoods.

## How Telecentres can Promote Livelihoods

Telecentres are public facilities where people can access the Internet, computers and other information and communication technologies to gather information, communicate with others and develop digital skills (telecentre.org). A telecentre may be, for example, a public library providing Internet access and basic digital literacy training courses (e.g. Biblioredes in Chile) or a community development centre providing, among other services, Internet access to meet the information needs of the community. Cybercafes are "privately owned, primarily urban establishment providing limited services, such as emailing and browsing". Telecentres may come in varied forms and names (e.g. village computing, information kiosks, community multimedia centres), but they have a common goal: to serve the community and support local development. This developmental characteristic distinguishes telecentres from cybercafes. Telecentres are often a key policy or programme to bridge the digital divide, and some Governments include a telecentre programme as part of their national ICT policy

The first telecentre programmes in developing countries started in the late 1990s. Financial, political and sociocultural sustainability concerns pose a challenge to the existence of telecentres and some of them have had to be discontinued or reduced in scale. Nevertheless, telecentres continue to play a critical role in supporting an inclusive society. Given the limitations of the private sector as regards serving less profitable areas, telecentres are increasingly recognized as a public good and a development instrument worthy of public support. Telecentres, since they support underserved communities, will often necessarily require financial subsidies. Governments' financial support for telecentres does not need to be permanent. In the longer term, telecentres may be able or expected to be self-sustainable, and the private sector may start offering some of the services in an affordable manner. Moreover, today, telecentres often benefit from an enhanced environment in which to operate and from a greater understanding of best practices for dealing with sustainability. Although there is no one sustainable telecentre model, sustainability can be achieved through the provision of value-added services, by working in partnership with others in the public and private spheres, and by timely scaling-up to build on existing experiences, capacities and resources.

How can telecentres support the livelihoods of people living in poverty? Telecentres may, for example, support the development of technical and business skills, provide access to key information, facilitate access to government services and financial resources, and/or provide micro-entrepreneurship support.

A study by Parkinson and Ramírez has assessed to what extent the Aguablanca telecentre in Cali, Colombia, is supporting livelihoods. It shows that the telecentre supported only some of the livelihood strategies employed by the community. Residents used the telecentre mainly to develop social assets(social relationships) and to increase financial assets in the long term by investing in education to improve formal employment prospects, even when current opportunities in the community lie primarily in informal employment. Other livelihood strategies were not being served through the telecentre, for example (a) increasing financial assets in the short term through the conversion of human or physical capital (i.e. formal employment or informal employment), (b) reducing reliance on financial assets by developing other assets (i.e. house ownership), and (c) reducing the necessary recourse to financial expenditure. The unemployed "rarely used the telecentre for searching a job and most considered it inappropriate for that use". The self-employed "rarely used the telecentre, or had any ambition to use it, in support of their business needs".

Telecentre managers and promoters can conduct similar analysis to better understand how specific telecentres can support local livelihoods. Such analysis, although context-specific, can help inform broader strategic decisions regarding the role of telecentre networks and support policymakers in their analysis and promotion of policies that can support the livelihoods of people living in poverty.

The following section provides a broader analysis of how telecentres, as institutions, are supporting the livelihoods of people living in poverty.

The analysis is based on key elements of the sustainable livelihoods approach, which includes the broader definition of poverty, an acknowledgement of people's different assets and the strategies women and men can follow, and a holistic approach (where in addition to building capacities, development efforts take the context, structures and processes into account). However, UNCTAD's analysis focuses on how telecentres support economic activities, and does not explore other areas important for sustaining livelihoods, such as physical well-being, which are not UNCTAD's areas of expertise.

### Telecentres' Impact on Supporting Economic Opportunities

To examine, at the broader level, to what extent current telecentres are supporting livelihoods, UNCTAD, in collaboration with telecentre.org, carried out a survey among telecentre networks on the impact of telecentres on promoting economic opportunities. The questionnaire targets telecentre networks rather than individual telecentres for a number of reasons. One reason, in addition to practical reasons, is that telecentre networks play an important role in creating dynamics within the telecentre movement, in pooling resources to, for example, develop content and training materials, in sharing best practices, in creating partnerships and, in summary, in being able to bring about changes.

The questionnaire was sent to 84 coordinators and leaders of telecentres, and over a quarter of them (22) responded. The respondents represented 22 different networks in 21 countries across Africa (7 responses), America (6 responses), Asia (7 responses) and Europe (2 responses), which serve over 7 million users a year. The large majority of the respondents were from telecentre networks in developing countries. There were four responses from two developed countries (Canada and Spain), and these are also valuable since those networks also aim at supporting vulnerable communities.

The following subsections present and review the survey's findings. The survey should not be regarded as a statistical representative account but as an overview of how telecentres are supporting economic activities in different circumstances. The results are contrasted with the existing literature on telecentres and complemented with information provided by the telecentres themselves. As the responses to the questionnaire come from diverse telecentre networks with different objectives, sizes, and experiences, and while the analysis takes into account the spectrum of networks, some of the results may not accurately reflect individual experiences.

***Profile of telecentre networks that responded to the questionnaire***

In terms of size, most networks (73 per cent) have fewer than 100 telecentres and a third have fewer than 25 telecentres but three large networks each have between 300 and 900 telecentres. Most of the networks are young: over half of them are less than five years old (established in 2003 or after), and over 80 per cent of them were established after the year 2000. In 2006 six networks were established; furthermore, there are indications that at least two other networks are being established in Rwanda and the Dominican Republic, and that discussions are taking place about setting up one in Indonesia. All of this suggests that there is an ongoing expansion in the establishment of telecentre networks.

In terms of budget size, nearly half of the networks reported that the annual cost of running the telecentre network was below $50,000 while for five networks the annual cost was between $250,000 and $5 million, and for four networks it was above $5 million. The networks generally diversify their sources of finance. Only one third of the networks are predominantly funded (over 75 per cent of total funds) by one source of finance, and when that is the case, they largely depend on government grants (three networks) or donor funds (three networks). Over two thirds of the networks charge user fees or charge for the sale of services to finance their activities. However, for most of those networks, user fees or the sale of services represent less than a quarter of all financial sources. Additionally, survey findings show that more than 60 per cent of the networks use four or more different sources of finance. There is only one exception: one network is entirely funded by government grants/subsidies. On average, donors and Governments are the largest providers of funds (providing, respectively, 30 per cent and 24 per cent of the funds of a network), followed by user fees, sales and in-kind contributions, which provide around 15 per cent each.

The telecentres served through these networks are mainly rural and multi-purpose (that is, telecentres, in addition to access to telephones, computers, the Internet and/or radios, offer access to other value-added services such as training and business support services).

In nearly a third of the networks there is a balanced representation of female staff – that is to say, women represent from 40 to 60 per cent of the network staff. However, one third of the networks still have a limited female ratio (women represent less that 30 per cent of the staff).

***Users***

To understand how telecentres support livelihoods, the first question is as follows: Whose livelihoods do telecentres support?

The 17 networks that reported data support over 7 million users per year. Half of the networks have fewer than 15,000 users per year, while the other half have over 100,000 users per year. The survey asked the telecentre network leaders to identify their two or three main user groups. Responses to the open question varied widely: some networks highlighted students, others the unemployed or new immigrants (in developed countries), but most responses did not identify the two or three main group of users.

Regarding female users, on average, women represent around 40 per cent of users, and the majority of telecentre networks have a balanced ratio (40 to 60 per cent) of male and female users. However, there are great differences in female user ratios across networks (ranging from 15 per cent to 60 per cent) and three networks have a female user ratio of below 30 per cent. Other studies also highlight significant variations in the gender balance among telecentre users to the detriment of female users.

It should be noted that the number and the gender of direct users do not necessarily reflect the total number and the gender of beneficiaries. For instance, a study in the Solomon Islands indicates that women often send their husbands to the radio station to transmit a message that they (the women) want to send.

Who uses and who benefits from telecentres largely depends on the network's aims and design. For example, in the telecentre of the Ajb'atz' Enlace Quiché association in Guatemala, an association that focuses on supporting education and training by and for Mayan people, three quarters of users come from an indigenous group (a reflection of the wider community) and the large majority of users are students, and among adults, nearly half of them are teachers. User demographic analyses reveal that all-purpose telecentres (telecentres that not specifically target one group of users) are not necessarily used by the whole community. For example, a recent survey shows that users from Nepal's Wireless Networking are mainly young (83 per cent were under the age of 30), male (72 per cent) and educated (all users are literate, although the national literacy rate is only 53 per cent).

### *Services provided*

A second step to understanding how telecentres support livelihoods is to examine the services offered by them. What range of services do telecentres offer? Do they provide specific services to support the economic activities of the community?

*General services*

A review of the types of technologies available across the telecentres reveals the kind, and delivery format, of services that can be provided. For instance, telecentres offering broadband Internet access can provide a wider number of services (e.g. downloading online training materials, VoIP) and perform a broader range of activities (e.g. buying and selling online) than telecentres offering limited access to the Internet.

Over 80 per cent of the networks provide access to computers in more than 75 per cent of their telecentres, that indicates that providing access to a computer is an essential service All the networks that responded to the questionnaire provide access, at least in some of their telecentres, to the Internet via dial-up or broadband, or both. However, broadband access to the Internet is still not available in a quarter of the networks. Few networks offer access to radio broadcasting, and when they do, it is to a limited extent – only one network offers it in more than 25 per cent of its telecentres. Access to telephone, fax and photocopier services varied widely across networks.

The range of services provided by telecentres evolves with technological developments and as different technologies become more affordable. For example, the telecentre network in the United Republic of Tanzania reports that telephone services are no longer offered because of the emergence of mobile telephones and an increasing use of VoIP.

*Training services*

Training services can help develop competencies that telecentre users need in order to conduct economic activities – that is, to enhance their human capital, which they may be able to exchange in the short or longer term for financial assets. Training, to be relevant, needs to build the competencies required for the economic activities that trainees undertake or will undertake and, additionally, trainees must be able to apply those skills. The questionnaire enquired about the different types of training services the telecentres provided and to what extent they were provided across the network.

Telecentres mainly focus on providing basic ICT skills: all networks provide basic ICT skills and three quarters of them consistently provide such training through their telecentres. More advanced ICT skills training in the use of advanced or sector-specific ICT tools and in the use of advanced functions of generic ICT tools is provided in nearly all of the networks, but a large proportion of them provide it only to a limited extent (i.e. in less than 25 per cent of their telecentres).

The development of basic literacy skills (the ability to read and write) is supported by numerous telecentres. Half of the networks provide training in those skills either consistently (i.e. in more than 75 per cent of the telecentres) or in a considerable number (25 to 75 per cent) of telecentres. This is one indication that networks are reaching people living in poverty.

Training to develop skills important for developing economic activities, including e-business skills,7 general business skills (e.g. marketing and management) and occupation-specific skills (e.g. farming, crafts and tourism), is provided in over half of the networks to a limited extent (less than 25 per cent of their telecentres). This means that there is scope for expanding the provision of that type of training by replicating within networks services already being provided by telecentres in the same network. Only one or two networks consistently provide (i.e. in more than 75 per cent of their telecentres) training in e-business or broader business skills. Some networks do not provide training services for e-business skills (three networks), general business skills (six networks) and occupation-specific skills (three networks).

What type of business-related training are telecentres providing? The telecentre network of Asturias (Spain) organizes training courses for businesses and entrepreneurs to become familiar with searching for employment opportunities and recruiting personnel, keeping financial records, conducting business communications, searching online for products, services and competitors, managing security (how to make security copies and protect from cybercrime), obtaining digital certificates, digitalizing documents and organizing business trips. After those training courses, the telecentres noted a significant increase in the number of wage employees and self-employed users. In Bangladesh, the Community Information Centres organize a training programme on "how to start up a new business", which explores information available on the web and how the telecentre can help users request a trade licence or a bank loan.

*Business-related services*

To further examine the extent to which telecentres support economic activities, the questionnaire enquired about the business support services being offered by telecentres.

The service most consistently provided, and by far, is "searching for information". All networks provide this service and 13 networks do so in the majority of their telecentres. The other two business-supporting services more widely available are "searching for/advertising jobs" and providing access to "government services" (a third of the networks provide those services in the majority of their telecentres). Other services often supported are searching for "professional/sector-specific information", "typing", "business communications" and supporting "employment opportunities". Some networks provide support for "designing and creating websites", "developing content", "buying and selling", and developing "business opportunities".

In Indonesia, to empower and mobilize poor communities for economic activities, and on the basis of the findings of ethnographic research, each telecentre has an infomobilizer – that is, a person that supports the development of the community by, among other things, using and promoting the use of relevant information. The infomobilizer helps the community/ village identify its needs and its opportunities to improve livelihoods (for example, acquiring new agricultural skills, expanding the marketing of village products and linking to experts for information). Moreover, websites to promote village products and tourism are developed.

Among responding telecentres, there is limited support in the areas of "advertising", "accountancy", "banking", "microfmance", supporting the carrying out of "payments", "export/import and trade facilitation", "data management and storage", "taxation", and "innovation/research and development". Each of those services is provided in less than half of the networks.

Successful experiences of other networks in supporting some of these services suggest that there is substantial scope for expansion. However, expanding the offer of business-related services requires the existence of certain conditions. For instance, supporting tax filing or other government services may need to be led by the Government. Also, some of the above areas (e.g. trade facilitation and banking) require specialized skills or additional infrastructure not always available.

Business-related services can be supported in a number of ways:

(a) Through specific training courses;

(b) As part of broader training courses – for example, a general course on ICT may show how to search the Internet for general and sector-specific information, conduct payments online and access different e-government services;

(c) Adhoc support provided by staff;

(d) Provision of the service by the telecentre itself.

Survey responses show that to support business-related services telecentres mostly provide the service itself or provide adhoc support.

Those services are supported to a lesser extent with specific training or as part of other training courses. Specific training is mostly offered for "searching information", "creating websites", and "typing", and secondarily for supporting "job searching" and "employment opportunities", for conducting "business communications" and accessing "professional or occupational specific information". No telecentre network offers training in "banking", or "export/ import and trade facilitation", and only one network offers training (as part of a broader training course) in conducting payments online.

Overall, there is limited explicit effort to build capacities in several business-related areas and to mainstream business-related training, the reasons probably being the greater complexity (telecentres have started by providing more basic services) and the specific knowledge requirements. Telecentre staff can provide adhoc support in a convenient manner, but it is unrealistic to expect that they can provide in-depth support in all the different areas. Rather, telecentres "in the future will not be a one-stop shop that requires a 'superman' office person, but a place where computing is a means for delivery of services from many back ends that support individual retail outlets". Some basic e-business training could be developed by the telecentre networks or integrated into more general ICT training courses (e.g. how to buy and sell online, online payments, etc.). Several telecentres are starting to include a wider range of specialized training services for micro-entrepreneurs. In Guatemala, Enlace Quiché, in collaboration with Fundación

Omar Dengo, is developing a training course for small and medium-sized enterprises (SMEs) in Central America to support productivity, facilitate their administrative and productive processes and support the development of management capacities, including leadership, entrepreneurship and technical skills. Training of greater complexity, particularly when it deals with specialized topics (e.g. exporting and importing goods), may be better developed and provided in collaboration with specialized organizations such as trade-supporting organisations, business-promoting agencies, banks, and so forth.

Often, as respondents concede, telecentres would like to offer business-related services, but they are limited by their "lack of resources and partners to support such projects" (Kenneth Chelimo, National Coordinator of Kenya's Network of Telecentres), and "the lack of information available in digital formats and the low levels of local business development and connectivity" (questionnaire respondent). There an expectation ("this is an area we have just started working in", Julia Pieruzzi, Uruguay National Director, CDI) that as networks acquire experience and as Governments develop online services, a wider range of business support services will be provided across the telecentres.

***Economic focus: Objectives and sectors***

While most of the telecentre networks that have responded to the questionnaire support economic activities where possible, less than a quarter of the networks reported that supporting economic activities is their main objective. Two telecentre networks (Red Conecta and Fundación CTIC in Spain) have a clear mandate that does not include supporting economic opportunities beyond supporting employment access through the development of ICT skills and job advertising. Several telecentres stress the importance of their social and educational goals, and the relevance of services in those areas for developing economic opportunities. Thus, as supporting economic activities is not the main goal for many organizations, resources are often not focused in these areas.

Telecentres have the potential to support different economic areas. For example, in the aboriginal northern communities of Canada, the Community Access Programme of Nunavut helps artists in the booming local crafts industry to create websites, to sell their products and to trade through eBay. Users in the tourism sector use the centres to create websites for advertising purposes. As the Secretary-Treasurer of N-CAP explained, local business people who cannot afford a computer and visiting scientists needing connection to the Internet use the facilities to gather information and manage e-mails. The telecentres also support visiting scientific researchers in various ways and with some specific projects. Many telecentres offer basic literacy training and access to distance education, as well as access to both government and non-government health information. Several of the telecentres specialize in film production and editing.

The survey asked each telecentre network whether their telecentres provide support or offer services related to twelve broad economic sectors. The two economic sectors in which telecentre networks provide most support or offer most related-services are, obviously, the information and communication sector, and the educational sector. Those sectors are consistently supported or serviced by over half of the telecentres. Another two widely supported sectors are arts and entertainment and the primarysector (agriculture, fishing etc.), which are consistendy supported or serviced in, respectively, a third and a fifth of the telecentres. The health sector is supported in nearly all of the networks; however, only two networks support this sector in the majority of their telecentres. Rather surprising is the fact that only two networks consistently support the public administration sector and half of the networks still do not support it. The tourism sector is supported by over two thirds of the networks, but no network consistently supports that sector. Trade and tourism are supported to a limited extent in 40 per cent of the networks, a fact which indicates that there is potential for sharing practices among telecentres within the same network. The manufacturing sector and professional and scientific activities are rarely supported, and services or support regarding the financial services sector are scarcely provided. In general, the responses from the survey indicate that telecentres support some economic sectors, but there is potential to intensify the support and offer related services for a number of sectors, such as public administration, trade and tourism, and to explore opportunities for telecentres to provide services important for conducting business, such as financial services.

To understand to what extent people living in poverty benefit from the services offered by telecentres, the survey enquired how often, and how, services are targeted to specific vulnerable groups (women, the unemployed, the disabled, people living below the national poverty line), to businessmen/

women, and towards specific occupations. Since several reports often suggest that a key user group is students, the UNCTAD survey also enquired about that group. Results confirm that students are the group most often targeted, as well as women and people living in poverty. Over half of the networks always target their services to people living below the national poverty line, women and students. And over one third of the networks always target their services to the unemployed and to specific occupations. The groups that are less often targeted are disabled people and entrepreneurs.

An overwhelming majority (over 90 per cent) of the telecentre networks indicated that each of the following methods of targeting services is employed in their network:

The training content is adapted to the target group.

- Courses take into account the needs of specific groups when designing the training (i.e. timing of the courses, location, etc).
- Some courses are specifically designed for the target audience.

However, only some networks (less than a third) offer financial support to target their services. Other methods of targeting services indicated by respondents, and not mentioned above, include targeted marketing and the fact that the courses are free for the users.

The only obvious conclusion that can be drawn from those responses is that telecentres use a variety of methods to target their services, but not financial support. Telecentre networks would need to conduct more specific and local analyses to be able to assess how effective the different methods are in targeting different community groups.

***The environment: Important factors and challenges***

The context in which telecentres work shapes their role and ability to support livelihoods. For example, the regulatory environment and the quality of the general infrastructure condition the type of technologies that are available in a community and their affordability. The level of development of the private sector, the public sector and civil society conditions the range and type of services that are available and can be offered through the telecentres.

Telecentre features (such as the objectives of the telecentres, their location, the range of services offered and their affordability) and the availability of relevant content are considered highly important by telecentre leaders – which is logical given their closeness to them. More importantly, respondents highlight the lack of relevant content as a condition not being met to support livelihoods, followed then by the range of services offered by telecentres and affordability. The fact that telecentre leaders consider that there is a greater gap in the provision of relevant content than in the affordability of telecentres reinforces the observation that "the content and services sector supporting village computing is still in its infancy" (Bell, 2006, p. 24) and the importance of having locally developed content, which is of good quality, and relevant to the livelihoods of the users. "Content and services that are particularly relevant and that have an increased potential to generate revenue are those that are integrated into business and trading activities".

Most telecentre networks receive some form of government support. However, public financial support is not available for telecentres in poorer countries (e.g. Bangladesh, Congo, Kenya, Nepal, Mali, United Republic of Tanzania). The ministries that most often provide support are the Mnistry of Telecommunications or ICT, followed by the Mnistry of Education, the Mnistry of Social Affairs/National Development and the Mnistry of Agriculture.

Among respondents, Chile – a State with a strong government vision to promote and adopt ICT – is the country where most ministries are involved in supporting telecentre networks. For instance, only respondents from Chile acknowledge receiving support from the Mnistry of Finance/ Economy/Trade.The Partnerships for e-Prosperity for the Poor (Pe-PP) in Indonesia is a programme funded by the United Nations Development Programme (UNDP) and implemented by the National Development Planning Agency. The national Government provides strategic support, involving many ministries, such as the Ministry of Communication and Information, the Ministry of Research and Technology, the Ministry of Education and the Ministry of Agriculture. It also collaborates by sharing content, training programmes, and networking telecentres and by funding buildings, operational costs (that is, Internet connection) and human resources. One of the objectives of the Pe-PP is to mainstream ICT for poverty reduction in the national poverty reduction programme

***Use and impact***

To understand the impact of telecentres, the survey asked telecentre leaders about the purposes for which the different user groups use the telecentre, and about their perception of the positive and negative impact that telecentres may be having in the community.

In terms of use, the responses were very clear. The three more common purposes of using a telecentre are "personal communications", "searching for information" and "receiving training". Other purposes, including "buying and selling goods and services", "business communications" and solving "administrative matters", are rarely perceived as one of the two key purposes. This is consistent with other findings. A study of Chilean telecentres (SUBTEL 2005) shows that according to users, telecentres' main impact on their quality of life was in the area of information and knowledge, as well as communication. Another study, this time of telecentres in five African countries (Etta and Parvyn-Wamaliu 2003), noted that they were mainly used for communication and entertainment, rather than for economic purposes. More specific analysis will be necessary in order to understand why in some cases telecentres mainly support informational and entertainment activities (Is it because of a lack of services available and a lack of support for conducting economic activities? Is it because there are other local organizations better placed to support such activities?), and to explore the scope and mechanisms for promoting a more varied and deeper use of telecentres.

By user group, the perception is that self-employed users make more diverse use of the telecentre, and that for that group "business communications" are more important than "personal communications". Other groups perceived as having, to a limited extent, other key purposes besides "personal communication", "searching for information" and "training" are the unemployed ("solving administrative matters"), the employed ("solving administrative matters" and "conducting business communications"), men ("conducting business communications"), and people living in poverty ("selling goods and services" and "solving administrative matters"). People engaged in family duties, students, women and minorities are perceived as having only three key purposes (the above three common ones) for using the telecentre. The perceived differences between use by men and use by women are that (a) men may have as a key purpose to conduct business communications, but not women; and (b) women are more interested in receiving training than men.

These results are based on the perceptions of telecentre leaders, and may differ from the actual situation. Being aware of those perceptions is valuable because perceptions guide the decisions that telecentre leaders make. However, all telecentre managers should corroborate perceptions with actual information, and telecentre networks should monitor on a regular basis the purposes for which different groups use the telecentre.

To understand what type of impact telecentres have in supporting the economic activities of their users, the UNCTAD questionnaire asked telecentre leaders to what extent telecentre networks help users to:

- Acquire new skills;
- Support existing economic activities;
- Develop new economic activities;
- Improve self-employment opportunities;
- Improve salaried employment opportunities.

According to telecentre leaders, the greatest impact comes from the acquisition of new skills, followed by improvement in self-employment opportunities. Over three quarters of the respondents are confident that telecentres are helping users acquire new skills and nearly half of them indicated that telecentres are improving self-employment opportunities. All respondents stated that telecentres are supporting existing economic opportunities, while three of them indicated that telecentres are not supporting the development of new economic opportunities. There are substantial differences in the perception of the extent to which telecentres improve salaried employment opportunities: according to nine respondents, telecentres are improving salaried employment opportunities, while for five respondents they do not. That difference can be partially explained by the context in which the telecentre operates: telecentres operating in more developed economies can better support salaried (often formal) employment opportunities, while telecentres operating in informal economies will have fewer opportunities to provide access to salaried employment opportunities.

Participants were also asked about changes observed in the lives of telecentre users, including changes in income levels and distribution, quality of life, access to public goods and services, coverage of basic needs (i.e. housing, health, nutrition), consumption, social relations and confidence levels. Most respondents highlighted improvements in confidence and changes in behaviour regarding the use of ICT Half of the respondents cited improvements in income levels and employment, as well as in skills development. Seven respondents cited improvements in the coverage of information needs, and four mentioned improved access to public goods and services. For example, the Grameenphone Community Information Centres provide income opportunities for the entrepreneur managing the telecentre, who is able to recover his investment in one year, and for the end-user, who gets a fairer deal in economic transactions where the middleman is eliminated. With increased income, people can afford basic necessities such as food, shelter, health facilities and education. Also, information on public goods and services is now available, and access to digitized forms used by the Government has been enhanced. There is also the possibility of accessing information on housing opportunities (real estate, house-building loans) as well as health information and services.

Most respondents did not indicate any negative impact of telecentres in the community. Those that indicate such an impact highlighted access to pornographic sites and cybercrime and "creating another area of necessary spending" (Joseph Sekiku, Interim Chairman, Tanzania Telecentre Network). A more precise analysis could identify other less visible negative impacts that may have important implications for broader community development, including further marginalization of non-users where "the learning obtained by those closest to the Telecentre in itself becomes another expression of power and control that interferes with the participation of those in the community that need it most".

## Telecentre Management

As telecentre networks are too often short of financial and human resources, and their capabilities rest on the ability to work in partnership with other organizations and to leverage support, an examination of the organizations that telecentres work with provides a good indication of the areas in which telecentres can support services. The findings of the survey show that over two thirds of the networks work regularly (that is, often or always) with social and women's organizations. Over half of the networks regularly work with the regional or local government and with secondary schools. Over a third of the networks work regularly with the employment office/a recruitment agency, but only two networks always do so. Only around a quarter of the networks work regularly with primary schools, universities, professional associations, business-supporting organizations, private companies and innovation organizations. And only a very limited number (or none at all) work regularly with trade promotion agencies, microcredit organizations, the tax office or investment promotion agencies.In general, telecentre networks work with social and educational institutions and, to a lesser extent, with organizations that promote economic activities (such as professional associations or business-supporting organizations). Therefore, there is scope for working with the latter organizations (for example, chambers of commerce or Trade Points) in order to, for instance, share/ provide training programmes and business-related services.

However, where the local and regional development of economic organizations is very limited – for instance, a respondent indicated that "some of the above organizations are hardly present at the regional even national level" – it will be unrealistic to expect telecentre networks that have no specific objectives with regard to supporting economic opportunities and that are not embedded in an economic activity to successfully provide services in those areas.

Evaluations, assessments and monitoring reports help understand what are the needs of a community, how well a telecentre is performing and who is using the service. They are important in designing and upgrading the activities of the telecentres. Moreover, they provide an indication of telecentres' objectives and management style and, in particular, of how they are supporting economic activities. The UNCTAD survey asked telecentre networks whether they had undertaken any study in any of the following areas:

- Needs assessment (an assessment of the needs of the community);
- Monitoring (assessing who uses the telecentres and for what);
- Evaluation (a study of the effectiveness, sustainability and impact of the telecentres);
- Livelihoods analysis (understanding the livelihoods of a community).

The large majority of networks have conducted an assessment of the needs of the community and have monitored who uses the telecentres, and for what. A smaller number of telecentre networks, although still the majority, have undertaken an evaluation or a livelihoods analysis. For instance, the Grameenphone Community Information Centre in Bangladesh has conducted several studies to understand the economic sector, business patterns, information gaps, what are the sustainability factors, which services are in demand, and what is the contribution to livelihoods, and it has a monitoring system.

Several networks conduct studies on an informal basis (for example, they are conducted by students or at the local level). Only five networks reported having the studies disaggregated by gender, income/poverty levels, education levels and/or occupation, and one respondent highlighted the difficulties in having data disaggregated by income levels. Those results raise questions about the ability of telecentre networks to target their activities without disaggregated data. Thus, as suggested earlier, it is important that monitoring activities and evaluations disaggregate data by gender, education, occupation and income/poverty levels. Telecentre networks can prepare some guidelines on how to collect disaggregated data, particularly by income/poverty levels, using some of the tools used by other developmental programmes and the telecentre's collective experience.

Telecentre networks were asked to describe the results of their studies. The responses were too limited, but some of the highlights were as follows:

- Networks located in developed countries, where the formal economy is more developed, noted that telecentres have provided employment-related skills (ICT skills, job searching).
- Community needs are great and the capacities of telecentres limited; thus, it is important to work with others.
- Two networks (in two developing countries) reported that beneficiaries are younger, educated males, while in more advanced economies they where reported to be unemployed aboriginals, and/or low-income rural dwellers and people with disabilities.
- There are difficulties in capturing the specific contribution of telecentres in terms of improvement of livelihoods, including limited resources to capture that impact at the household level.

Without more information on the outcomes of the evaluations carried out, it is not possible to assess the contributions that such evaluations make to understanding how telecentres support livelihoods. Nevertheless, the responses show that telecentres have different capacities with regard to undertaking monitoring and evaluation activities that affect the ability to target services and thus to have an impact on the livelihoods of men and women.

## The Future

The survey asked telecentre network leaders to identify three areas in which they would like to receive more support. The two areas selected most often were: support for developing the skills of telecentre staff, and support for accessing and developing relevant content. The latter was consistent with earlier responses that identify the availability of content as an important condition not being met. Half of the leaders would also like to receive support for ensuring the sustainability of telecentres and delivering a wider range of services. Eight respondents would also like to receive support for promoting entrepreneurship and business opportunities. Telecentres leaders are not particularly interested in receiving support for making services more affordable for users, a finding that is not consistent with the earlier finding that affordability is an important condition not fully met.

These responses tend to highlight the interest of telecentres in providing a more in-depth and quality service, rather than in expanding the telecentre network. The sections that follow provide some suggestions about how to support access to relevant content, deliver a wider range of services, and promote entrepreneurship and business opportunities.

Telecentre network leaders believe that the institutions that should be further involved in improving livelihoods are primarily civil society organizations and local governments . Respondents gave a lower priority to the involvement of the private sector and the national Government. This result is somewhat inconsistent with earlier responses stating that the private sector is an important condition not being met. A possible explanation may be the lower levels of confidence regarding the likelihood of private sector involvement. In any case, a more detailed analysis would be necessary in order to ascertain what specific support may be required from different institutional actors for a given context. For example, in Chile, telecentres are strongly supported by the national Government, and stakeholders would like to see greater involvement of local governments and civil society organizations.

In summary, the survey's findings show that telecentre networks come in different sizes and formats but are mostly rural and multi-purpose, providing access to a range of services. Computer access is widely offered, while access to the Internet, particularly broadband, is more limited. Most telecentres provide training in basic ICT skills, few provide advanced ICT skills training, and a number of networks consistently support the development of basic literacy skills. Training to develop general business, specific e-business and occupational skills is provided consistently in a small number of networks and to a limited extent in others. The business-related services most often supported, generally through telecentres providing the service themselves or offering adhoc support, are searching for information, searching for employment and access to government services. Specific training in business-related services is mostly offered for searching for information, creating websites and typing. The three key purposes of using a telecentre are searching for information, personal communications and receiving training.

The majority of networks support economic activities where possible, but this is not their main objective. The main economic sectors serviced by telecentres are the ICT and the educational sectors. With regard to supporting livelihoods, the single most important factor that needs attention is the availability of relevant content. Other key factors requiring support are the quality of the general infrastructure and of economic and business conditions, as well as the development of a critical mass of ICT users, of a wider range of services and of occupational and business skills. There is also scope for greater involvement of the private sector, in particular business-supporting organizations, and civil society, in that order.

## Best Practices and Opportunities

On the basis of the above findings, as well as available literature on telecentres, this section provides an overview of best practices for supporting livelihoods, particularly economic activities, through telecentres. The approaches highlighted may not necessarily apply to all telecentre contexts but provide an indication of how livelihoods can be better supported.

### Facilitating Livelihood Strategies

Successful innovations and business propositions are those that make life easier for users and customers. Telecentres can make life easier and provide value added for their users by supporting their economic activities and livelihoods strategies. To do so, they must focus on the provision of services rather than on the provision of connectivity.

Value can be created by providing facilitated access to information and to more or enhanced government services, and by allowing transactions. By providing access to government services such as land records in India offering customized services (such as specific agricultural information in the Pallitathya Kendra in Bangladesh) or saving time spent in acquiring market information in rural Nepal, telecentres make life easier. The services that provide value added depend on the context. Some may already be offered satisfactorily by other institutions or their provision may be determined by other institutions (such as, e-government services). "ICTs alone cannot improve the service delivery to rural poor. Significant re-engineering of backend processes and introduction of services that directly contribute to the poverty alleviation are needed to make such initiatives sustainable".

Telecentres unable to provide value-added services become irrelevant. For example, an evaluation of the Gyandoot telecentre network in India showed that the rural poor did not perceive the telecentre as a platform for seeking government services, because there were alternative and preferred ways for them to obtain those services.

Responses from UNCTAD questionnaire show that there is scope for providing a wider range of value-added services. For instance, the availability of training to develop skills important for undertaking economic activities (such as e-business skills) is still limited. To continue providing value, telecentres have to plan on providing a continuum of services, from basic ICT skills to more specialized training, and support customers in using those skills and trying out improvements in their daily activities. In summary, telecentres can offer additional value by providing e-business skills training, supporting the use of ICT for specific or sectoral activities, and facilitating access to markets, finance and knowledge relevant to the livelihoods of the community.

## Embedding ICT in economic activities

E-Choupal is one of India's most successful programmes using ICT to support the economic activities of people living in rural areas. Through "network orchestration" it caters for underserved rural markets and helps farmers halve transaction costs. E-Choupal is a commodity services programme that supports farmers through over 5,000 information kiosks providing real-time information on commodity prices, customized agricultural knowledge, a supply chain for farm inputs and a direct marketing channel for farm produce. Because the network is strongly embedded in a specific economic activity, it enables its participants to derive economic opportunities.

A similar project increasingly embedding ICTs in economic activities relevant to the poor – although not strictly a network of telecentres – is the Dairy Information Service Kiosks (DISK). DISK is a pilot programme that supports milk cooperatives in making better use of existing information on the quality, quantity and price of the milk created by automated machines of dairy cooperatives, as well as in improving access to information on dairying and milch cattle, as well as other services.

Both networks are based on market approaches supported by a larger organization (the private ITC and the Anand District Milk Cooperative). In a vertical integration model, the fact that there is a telecentre is secondary. The focus is rather on the exploitation of ICTs to support a specific economic activity, whether trading commodities or supplying milk. E-Choupal has developed a one-stop shop offering a wide range of services for farmers. The downside of telecentres embedded in one particular economic activity is that those not part of the activity will be excluded, and unless the services offered are expanded to support other areas, the benefits are limited to the specific economic activity.

None of the telecentres that responded to the UNCTAD questionnaire are strongly embedded in one specific economic activity. Except when an existing economic organization such as an occupational association (i.e. a fishermen's association) or a private business, launches it, telecentres follow a more diversified strategy for supporting economic opportunities. In this case, providing value-added services and developing niches of economic opportunities are the two options to strongly support economic opportunities.

## Developing Niches of Economic Opportunity

One way of creating economic opportunities is to support clusters of economic activity. By developing support and knowledge in one area, a telecentre can, by virtue of concentrating resources and developing specialized know-how, provide additional economic opportunities.

Telecentres can play a role in developing clusters of economic opportunity by providing access to knowledge and opportunities for the development of expertise and relationships. For example, in Clyde River, a small impoverished community of 820 people in Nunavut (Canada), nearly two thirds of the adults have not finished high school and are not working in salaried positions nor are they self-employed. As Darlene Thompson, Secretary-Treasurer of N-CAP, explains, "following an increased interest in film production and edition, the telecentre looked for additional funding to purchase filming equipment and provide sector-specific training. As a result, there is a core group of young people trained in the industry, and film companies are taking an interest in using this community for making films given the availability of trained personnel." In response to growing demand from community members and visitors, and to an increasing number of job opportunities, the network has also developed a programme to support scientific research work, which offers services for visiting researchers (such as access to the Internet, rental of wireless modems, printers) and training in basic research methods, including logistical coordination and global positioning systems, for community members.

Other responses from the survey provide examples of economic sectors in which telecentres are working, but there is not much evidence about telecentres specializing in key economic sectors other than ICT and education. As indicated earlier, there is scope for expanding support to sectors such as trade and tourism. Developing niches of economic opportunities for the local community can help provide a reason for using the local telecentres and can thus help reinforce their sustainability. If specific know-how and valuable services were developed, users would be ready to pay more for them.

## Providing Specific Support

If telecentres are to support people living in poverty, market approaches alone will not suffice. Specific efforts are needed to support those in weaker positions. For instance, "the poor are under-represented in accessing the MSSRF knowledge centres in India, as information about crop production and market prices is of interest mainly to landowners and not to the poor, who are predominantly landless". Specific support to those that need it most can be in the form of infomediaries, specific programmes targeted to groups in a disadvantaged position and diversification of services to support the economic activities of the poorest.

Community infomediaries are the linkage between the knowledge and content available through ICTs and individuals. They are familiar with the Internet and ICT and help translate specific needs into information needs and solutions. In Indonesia, each telecentre of the Partnerships for e-Prosperity for the Poor has a manager, an IT administrator and a community development specialist (infomobilizer) that helps integrate "access to information and communication technology with community empowerment activities". In Bangladesh, D.Net works with infomediaries recruited locally and trained in basic ICT issues, documenting processes, mobilization and marketing the telecentre. It helps villagers ask the help desk questions about their livelihood, or they may themselves look for answers in the content database. Infomediaries are particularly important in communities with low literacy levels, strong preferences for face-to-face communications and greater needs for accompanied support.

Specific programmes targeted at disadvantaged community groups are necessary in order to support equality of benefits and prevent exclusion patterns from being exacerbated. For example, women are often at a greater disadvantage than men in benefiting from ICT, and thus specific efforts are needed. In addition to mainstreaming gender concerns in telecentres – by, inter alia, ensuring balanced participation in telecentre use and management – there is scope for providing specific training and services for women and for working with institutional mechanisms that are gender-sensitive, such as women's self-help groups.

**Supporting Access to Relevant Information and Knowledge**

Information and knowledge provide the foundation for economic development. Providing access to information and supporting the development of information and knowledge remain a key pillar for supporting livelihoods.

To facilitate access to relevant information some telecentre networks have focused on developing content for users. For instance, the Swaminathan Foundation's village computing project in India devotes specific efforts to creating, repackaging and disseminating content to its telecentres. The Manage cyber extension initiative in India has developed CD-based learning packages for, inter alia, making pickles, and women are now able to sell pickles in the local market. Other networks, such as the Grameenphone Community Information Centre (GPCIC), work with third-party content providers.

Users often require support and/or have specific needs for information and knowledge not readily available. To help meet those needs, several initiatives have put in place help desks. For example, in Chile, SERCOTEC, a public institution supporting SMEs with telecentres integrated into its support offices developed in 2002 an online advisory service enabling users to post online specific questions to advisers. The latter are committed to answering those questions within 48 hours. The questions and their answers, as well as users' evaluation of the latter, are posted online in order to promote a wider dissemination of knowledge and greater accountability. Collaboration agreements with other organizations (such as universities, government institutions, banks and business-related associations) have brought in experts able to advise a wider range of specialized subjects, including taxation, employment legislation and financial services.

## Using Experience to Advocate for an Enabling Environment

Telecentres often operate in less than ideal environments with inadequate regulations for conducting e-business or, more generally, inappropriate telecommunications regulatory environments and trade regimes. Telecentre networks can use their experience and advocate for policy changes. Networks in Nepal and India are advocating for changes in government policies to promote a more conducive environment enabling users to benefit from ICT and participate in the information economy.

In India, to operate in a particular state, e-Choupal needs to ensure that the Agricultural Produce Marketing Committee Act is reformed before it can purchase grains directly from the farmers. The Act requires that certain grains be purchased from the official middleman (mandi). ITC, the parent company, has successfully convinced policymakers in different states to amend or allow specific exemptions to the Act since farmers can be better served by e-Choupal.

More broadly, in Nepal, ENRD has lobbied the Government to deregulate the import and use of the industrial, scientific and medical bands, so as to make VoIP free, at least for calls from computer to computer and from computers to Nepal Telecom landline telephones. It has also argued for decreasing ISPs' license fees, so that small business entrepreneurs can start ISP companies also in rural areas, and for subsidizing in each district of Nepal an organization willing to establish a community internet service provider (CISP) and use wireless technology to connect remote villages.

## Developing a Holistic but Precise Understanding

To support economic opportunities, a holistic but precise understanding of community livelihoods and economic activities is essential. Most telecentres have developed some type of monitoring or assessment of their activities.

Networks that are able to fully understand the environment and community needs, as well as the impact of their activities, can confidendy provide relevant services. Understanding how ICT services support local economic activities is a first step. An evaluation of the Solomon Islands' People First network (PFnet), a rural connectivity project, showed that there is scope for promoting PFnet services for business activities that enable users to earn a livelihood, by creating awareness and training people in new ways of accessing information and opportunities".

Understanding who benefits and how from a telecentre programme is also indispensable for supporting the livelihoods of men and women living in poverty. For instance, field evidence from Uganda reveals fhat "the use of ICT facilities was and is male dominated with women often focusing on uses that integrate ICT in their already existing purposes more than for entrepreneurial activities. The latter uses are known to generate higher returns than the former".

Survey responses and the material provided indicate that telecentres often still have to develop a more precise understanding of the impact of their activities on different groups. For instance, six out of 23 respondents did not provide the percentage of female users. Consequently there is room for concern about how a telecentre can support the livelihoods of men and women without having basic data on who is using the telecentre.

Financial and human resources constraints are often a deterrent with regard to carrying out monitoring activities, assessments and analytical studies. However, several networks work with postgraduate students to carry out monitoring and evaluation studies. Similarly, there are universities that are carrying out studies on, and supporting, the work of telecentres.

## References

Amariles F, Paz O, Russell N and Johnson N (2006). The impacts of community telecentres in rural Colombia.

E-Choupal (2006). ITC e-Choupal presentation at UNCTAD's Expert Meeting on Enabling Small Commodity Producers in Developing Countries to Reach Global Markets, Geneva, 11–13 December 2006.

Etta F and Parvyn-Wamahiu S (eds.) (2003). *Information and Communication Technologies for Development in Africa, Vol 2, The Experience with Community Telecentres.* Available at www.idrc.ca.

UNCTAD (2007). UNCTAD questionnaire: promoting men's and women's livelihoods through ICTs. The case of telecentres. April 2007.

# Bibliography


Ahmad A (2006). *Management information systems (MIS) for microfinance.* Foundation for Development Cooperation, Brisbane, Australia.

Alistair Duff (2000) *Information Society Studies.* London: Routledge.

Antonelli, C, (1992), The economic theory of information networks. In: C. Antonelli (ed.), *The Economics of Information Networks*. North-Holland, Amsterdam., 5-27.

Babe, R. E., (1995), *Communication and the Transformation of Economics: Essays in Information, Public Policy and Political Economy.* Westview, Boulder, CO.

Baily MN (1986). Productivity growth and materials use in US manufacturing. *Quarterly Journal of Economics,* Vol. 101, No. 1, pp. 185–196.

BIS (2004). *Survey of Developments in Electronic Money and Internet and Mobile Payments.* Committee on Payment and Settlement Systems, Bank for International Setdements, March.

Carnoy, Martin. (2005), *ICT in Education: Possibilities and Challenges.* Universitat Oberta de Cataluny.

Carr N (2004). *Does IT Matter? Information Technology and the Corrosion of Competitive Advantage.* Harvard Business School Press.

Christian Fuchs (2008) *Internet and Society: Social Theory in the Information Age.* New York: Routledge.

Cracknell D (2004). Electronic banking for the poor: panacea, potential and pitfalls. *Small Enterprise Development,* vol. 15, no. 4, pp. 8-24.

Criscuolo C and Waldron K (2003). *E-Commerce and Productivity.* Economic Trends 600, UK Office for National Statistics.

David R. Smith, (2003).*Digital Transmission Systems*, Kluwer International Publishers.

E-Choupal (2006). ITC e-Choupal presentation at UNCTAD's Expert Meeting on Enabling Small Commodity Producers in Developing Countries to Reach Global Markets, Geneva, 11–13 December 2006.

Etta F and Parvyn-Wamahiu S (eds.) (2003). *Information and Communication Technologies for Development in Africa, Vol 2, The Experience with Community Telecentres.* Available at www.idrc.ca.

Frank Webster (2002) The Information Society Revisited. In: Lievrouw, Leah A./ Livingstone, Sonia (Eds.). *Handbook of New Media*. London: Sage. pp. 255–266.

Frank Webster (2002) *Theories of the Information Society*. London: Routledge.

Grossman, G. and E. Helpman (2005), "Outsourcing in a global economy", *Review of Economic Studies* 72: 135-159.

Hopwood, A. G. & P. Miller (eds), (1995), *Accounting as Social and Institutional Practice*. Cambridge University Press, Cambridge.

International Telecommunication Union. (2009). *Measuring the Information Society: The ICT Development Index*. pp. 108.

Irvine, Martin. (2006). "Postmodern Science Fiction and Cyberpunk", retrieved 2006-07-19.

John Proakis, (2000).*Digital Communications*, 4th edition, McGraw-Hill.

Lamberton, D. M. (ed.), (1974), *The Information Revolution. Annals of the American Academy of Political and Social Science,* 412.

Long, Paul; Wall, Tim (2009). *Media Studies: Text, Production and Context*. Pearson Education.

Markman, K. M. (2006). Computer-mediated conversation: The organization of talk in chat-based virtual team meetings. Dissertation Abstracts International, 67 (12A), 4388. (UMI No. 3244348)

Oliver, Ron. (2002). *The Role of ICT in Higher Education for the 21st Century: ICT as a Change Agent for Education*. University, Perth, Western Australia.

Sergio Benedetto, Ezio Biglieri, (2008).*Principles of Digital Transmission: With Wireless Applications*, Springer.

Sterling, Bruce. (1992). *The Hacker Crackdown: Law and Disorder On the Electronic Frontier.* Spectra Books.

Stiglitz, J., (1985), Information and economic analysis: a perspective. *Economic Journal,* 95, 21-41.

Temin, P. (ed.), (1991), *Inside the Business Enterprise: Historical Perspectives on the Use of Information.* University of Chicago Press, Chicago.

Thurlow, C., Lengel, L. & Tomic, A. (2004). Computer mediated communication: Social interaction and the internet. London: Sage.

Timothy Binkley (1988/89). "The Computer is Not A Medium", *Philosophic Exchange*. Reprinted in *EDB & kunstfag*, Rapport Nr. 48.

UNCTAD (2008). *The Impact of ICT Use in Manufacturing Firms in Thailand.* Joint UNCTAD-Thailand National Statistical Office project.

Walter Ong, (1988). *Orality and Literacy: The Technologizing of the Word.* London, UK: Routledge.

Walther, J. B. (1996). Computer-mediated communication: Impersonal, interpersonal, and hyperpersonal interaction. *Communication Research, 23*, 3-43.